# CONTRACTING FOR ENGINEERING AND CONSTRUCTION PROJECTS

# Contracting for Engineering and Construction Projects

**Fourth Edition**

**P.D.V. Marsh**

Gower

No responsibility for any loss whatsoever occasioned to any person acting or refraining from action as a result of the material contained in this publication can be accepted by the author or publishers.

First published 1969
Second edition 1981
Third edition 1988
This edition published by
Gower Publishing Limited
Gower House
Croft Road
Aldershot
Hampshire GU11 3HR
England

Gower
Old Post Road
Brookfield
Vermont 05036
USA

British Library Cataloguing in Publication Data
Marsh, P.D.V.
  Contracting for Engineering and
  Construction Projects. – 4Rev.ed
  I. Title
  624


ISBN 0–566–07628–4

Library of Congress Cataloging-in-Publication Data
Marsh, P.D.V.
  Contracting for engineering and construction projects /
  P.D.V. Marsh — 4th ed.
    p. cm.
  Includes index.
  ISBN 0–568–07628–4
  1. Engineering contracts. 2. Construction contracts. I. Title.
TA180.M25 1995
624'.068—dc20     95–13028
                CIP


Typeset in Great Britain by Bournemouth Colour Press and printed in Great Britain by Biddles Ltd, Guildford.

# Contents

## APPENDICES

# Illustrations

**Figures**

# Preface

The period since the last edition of *Contracting for Engineering and Construction Projects* was published in 1988 has been one of radical change both in the law and commercial practice relating to engineering and construction projects.

The application of the law of negligence has been severely restricted by the decision of the House of Lords in Murphy v Brentwood District Council, so that it now has little application in disputes arising between employers and sub-contractors or as between employers and subsequent purchasers. Although currently there are proposals for reform in this area, at present the employer is forced to protect himself by obtaining a multitude of collateral warranties.

The revision of the EC Directives on Public Supplies and Works and the introduction of the Utilities Directives and the Remedies Directives into English law has meant that much more attention has now to be paid within the public sector to compliance with their formal procedures for the inviting of tenders and the award of contracts.

The privatization of many previously nationalized or publicly-owned organizations has meant a more cost-conscious approach to contracting. This, coupled with a down-turn in demand within the construction industry has resulted in greater commercial pressures being applied on contractors which they have sought to pass on to their sub-contractors. Nor have consultants been immune from these pressures. There is a sharply increased emphasis on competitive tendering on fees often on a lump sum basis. Design and build contracts have become more widely used, even in civil engineering, with consulting engineers being employed by contractors.

There seems no doubt that adversarialism in the industry has increased and it was against this background that Sir Michael Latham was asked to investigate the procurement and contractual arrangements in the UK construction industry. His report *Constructing the Team* published in July 1994 made many suggestions as to how relationships in the industry and its competitiveness can be improved and the total costs of construction reduced. A number of his specific proposals are referred to in the text.

It is now becoming increasingly recognized that conditions of contract should provide a positive encouragement to good project management and that in this way many of the adversarial problems associated with the traditional forms of contract and sub-contract could be reduced substantially. The New Engineering Contract (NEC) developed under the auspices of the Institution of Civil Engineers, but intended for use on any single or multi-disciplinary projects, has this as its objective. Initial reactions to its use are encouraging and Sir Michael Latham has strongly recommended the introduction of a single form of contract based on the principles of the NEC. Other traditional forms of contract have been revised since the first edition. In particular MF/1 now into its third edition has replaced the old Model form A.

Finally the long established method of resolving disputes in the industry, initially by 'the engineer/architect' and subsequently by arbitration or the courts, is now widely rejected as being too slow, too expensive and as promoting adversarialism. The use of adjudication is a key feature of the NEC and has been strongly supported by Sir Michael Latham. It is expected that even with conventional forms of contract this will become the norm in the none too distant future.

All these developments have been taken account of in the text even though inevitably detailed consideration of those that are more speculative has had to be postponed, since it remains the purpose of the book to continue to provide a basic guide to the commercial issues in contracting as they exist today.

PM

# Table of Cases

# Part One
# CONTRACT PLANNING

# 1 Planning process

The business of contracting is the planned utilization of resources, engineering skill, labour, materials, time, and money in such a way that both parties to the contract receive the maximum benefit from the resources expended. This book is concerned with the commercial aspects of that business – the methods by which the employer plans his project and the contractor his tender, the tendering process itself, the placing of the contract, and finally its administration.

Because it is the employer who must initiate and set the stage on which the contractor performs, we start with the employer. Because before action there must be planning, we start with the planning stage in the employer's office.

Success in contracting starts with clear and realistic planning tailored to meet the requirements of the objective which management has established. Before the employer can get to the point of inviting tenders, or opening negotiations with a single contractor, there must be a plan for the contract itself, and where there is a composite project comprising a number of contracts, a contract plan for the project as a whole. It is easy for this planning stage to be overlooked or for it to be regarded as the prerogative of the engineer or architect. It is, however, not only a vital stage to which the most careful thought requires to be given, but also one with which the commercial department of the employer should be most closely concerned, for it is here that money and time can be most easily lost or saved. Plan before you start work – the first law of contracting.

Planning must proceed in logical steps. It is suggested that those for a constructional or engineering project are:

1   Identification of the *objective* to be achieved.
2   Expression of that objective in terms of *time*.
3   Assessment of the *resources* both available and required.
4   Establishment of the most economic *method* by which the objective can be achieved by the time required, taking into account the resources available.
5   From the method selected, the allocation of *responsibilities*.

## Planning team

But before any plan can be made there must be a planning team. So the first step management must take, having decided to initiate a project, is to appoint one. Initially that team should be internal to the company and specifically should not include any outside consultants or contractors. All too often time is wasted, misunderstandings occur, and the project gets off to the worst possible start because the first thought of management, faced with a problem, was to seek outside help with no clear idea of what they were to be appointed to do or how they would work. By all means let the team get ideas from consultants or contractors as to what they could do, how long it would take, what it would cost, and what responsibilities they would accept. These points can then be considered by the team and taken into account in the action which they recommend should be taken. But the time to bring in outside assistance is after the plan has shown:

1   That it is necessary to do so.

2 What type of assistance is required.
3 What function the outside firm are to be brought in to perform.

As to who should be in the team, this will obviously depend upon the size and complexity of the project. However, the following are thought to be essential:

1 The project manager.
2 The project engineer. On a very small project he may be the same person as the project manager, but ideally the two should be separate, as the project manager's function, as his title implies, is management, not engineering.
3 A representative of the department or function for whom the project is being constructed. Again this might be the project manager, but more probably would be someone from marketing, production, or operations, according to the nature of the project concerned.
4 The purchasing or contracts officer who will be responsible for providing the commercial input to the project manager, advising on the contract strategy to be used, inviting and receiving tenders and placing the contract.
5 A representative of the finance department. For a small project which the employer is going to fund himself this could be a project accountant. On a large project, and especially if the firm is looking for the project to be at least partially financed by external borrowing, then it should be the finance manager or treasurer. If international tenders are being invited and multi-currency borrowings are involved, then it would probably be advisable for the firm to engage the services of a merchant bank to act as their financial advisers.

## Definition of planning objective

Any purchase is almost of necessity a compromise. There are few occasions when any employer can afford to have the best of everything, even if this were obtainable. Shorter delivery may only be achieved at the expense of higher prices. What one can afford may determine the quality of what one can buy. Shortage of capital may cause the purchase of equipment with high maintenance costs. Shortage of labour, or the need to reduce dependence on labour, may necessitate the purchase of equipment with a substantial degree of built-in automation. The absolute need from the safety angle to ensure complete reliability and conformity with rigorous specifications may limit the choice of suppliers to those possessing the highest standards of quality control.

The process of defining the objective starts therefore with the selection of those factors which are regarded as being of the maximum importance to the transaction in question. Sometimes from even a cursory examination one factor will stand out as of vital significance. It may be time of delivery. Once this has been established, then all subsequent actions will need to be subordinated to its achievement: the selection of the supplier, the formulation of the specification, the placing and wording of the contract, the action on progressing; all must be compatible with the defined objective.

More often no single factor stands out so clearly that others can be ignored. Certainly delivery on time may be important, but so too may be quality and price. Some sacrifice may be necessary in the interest of speed, but there are limits beyond which the pursuit of speed may become largely a self-defeating exercise.

Having selected therefore those factors which require to be taken into account, the team must then evaluate each factor relative to the others. Just how much is the employer prepared to pay for speed? Is he, for instance, prepared to dispense with competitive quotations, assuming that this is permitted under the EC Procurement Directives or the Utilities Directives? (See further p. 83.) From this evaluation a pattern will emerge: 'Yes, there is time to go to tender, provided this can be done in three weeks. Completion is essential within six months from the date of order, and therefore the choice of tenderers must be restricted to firms who can be relied upon to meet that programme, and extra progressing will be required. The order must make this clear, and any tender hedging on firm delivery will not be considered. There is a price ceiling, but this could be raised 10 per cent if this would ensure prompt completion'.

In many instances these processes of thought are elementary and may be carried out quite simply in a few minutes. With other transactions, however, and in particular with a project where a series of individual orders/contracts are closely inter-related, the problem becomes more complex and the relative importance of the various factors that much more difficult to assess. Here it becomes important to examine each of the activities necessary to the achievement of the overall objective, and to define for each the function it has to perform in the scheme as a whole.

To the extent that certain factors will be in conflict one with another, and as a preliminary step to the analysis of tenders when these are received, it is suggested that in a case of any complexity a formalized decision technique should be used in order to enable the team to establish the criteria for the optimal choice. The technique proposed is that of riskless utility in which an index is constructed representing the team's preferences over the range of likely variables. For a more detailed review of the use of this technique see the author's *Contract Negotiation Handbook*, Gower, pp. 32–43.

The basic assumption behind the construction of any such index is that riskless utility is additive, i.e. the utility of any offer to the employer is the sum of its component parts: price, quality, delivery, etc. The method of construction of the index is as follows:

1  Decide on those factors which are most important to the employer, e.g. price and delivery.
2  Value each of the factors on a scale 0 to 1 according to the team's assessment of its worth to the employer.
3  Decide on a weighting factor to be used when combining the two factors together which represents the opinion of the team as to the relative importance of one factor in relation to the other.

4  Combine the two factors together to arrive at a weighted value for each anticipated combination.

The additive model is based on the following assumptions:

- the ordering of preferences is transitive
- the ordering of each of the sets of factors is independent of the other – the employer if he is concerned with delivery will always prefer the offer quoting the more favourable delivery if the prices quoted are the same
- the strength of the employer's preference for any one factor is not affected by the other factor with which it is paired, i.e. the value of say two months saved on delivery is the same regardless of the price level.

The combination having the highest weighted combined score then represents the employer's first choice. It should be noted that if any factor has a value which is below that which is acceptable to the employer, e.g. a price substantially in excess of the budget, then that combination is automatically assigned the value zero irrespective of the worth of the other component.

A simple numerical example to illustrate the use of the method is given in Table 1.1.

In practical terms the area figures in bold type in the example represent that series of combinations which the employer would be likely to find acceptable and therefore defines the objective. Delivery must not exceed 28 months and then only if the price is highly competitive; price must not exceed £1 200 000 irrespective of how favourable the delivery period.

While it may be argued that constructing a simple mathematical model of this type does no more than rationalize intuitive thinking, it is considered that the act of the team in sitting down in a group and working out together the values and weighting factors provides a most effective means of concentrating thought and compelling the team as a whole to arrive at a consensus.

Where the tender is being invited subject to either the EC Procurement Directives or the Utilities Directives then the

**Table 1.1 Simple numerical example of planning**

| FACTOR | | | | PRICE | | |
|---|---|---|---|---|---|---|
| | | *Price in £* | | | | *Utility* |
| | | 1 000 000 | | | | 1.0 |
| | | 1 100 000 | | | | 0.9 |
| | | 1 200 000 | | | | 0.7 |
| | | 1 400 000 | | | | 0.2 |
| | | | | DELIVERY | | |
| | | 24 months | | | | 1.0 |
| | | 26 months | | | | 0.8 |
| | | 28 months | | | | 0.5 |
| | | 30 months | | | | 0.2 |

COMBINATION MATRIX   Weighting factors   Price 0.7   Delivery 0.3

| Price £'000 | 24 | 26 | 28 | 30 | months |
|---|---|---|---|---|---|
| 1 000 | 1.0 | 0.94 | 0.85 | 0.76 | |
| 1 100 | 0.93 | 0.87 | 0.78 | 0.69 | |
| 1 200 | 0.79 | 0.73 | 0.64 | 0.55 | |
| 1 400 | 0.44 | 0.38 | 0.29 | 0.20 | |

establishment of the objective will be of great importance to the definition of the criteria for the award of the contract. (See further p. 84.)

Finally, unless the objective has been carefully defined from the outset difficulties will be encountered at a later stage. Any issues having a fundamental effect on cost, time or resources must be specified so that all concerned are aware of the constraints under which they are working.

# Time by which objective is to be achieved

The time available will have a significant effect on:

1 The method of contracting – that is, there may be no time to go to competition, or a method must be selected which puts the primary emphasis on time as opposed to cost.
2 The cost of working overtime.

3  The method of construction – for example, use of duplicate form work, sliding shuttering, etc.

Time must be closely related to the definition of the objective. 'Completed' does not necessarily have the same meaning as 'in commercial operation'. Indeed, there may be a gap of many months between those events on a complex chemical plant. So in writing the definition of the objective as related to time it is essential to be completely unambiguous. It is also important to be realistic. The dictates of higher management must not be allowed to override common sense, and the wish should not be permitted to become father to the thought. Any engineering contract or project consists of a number of closely related activities, some of which can proceed in parallel, others only in series unless a 'fast-tracking' method of contracting is used with its attendant risk of cost over-runs. (See further p. 41.) Whilst some acceleration of a normal programme is usually possible at a price and with the use of additional resources, there is accordingly a very definite practical limit to the amount of acceleration possible. Such a limit is a function only of the job itself.

## Assessment of resources

The three aspects to be considered here are:

1  The employer's design, engineering, and management resources available and how these match up to those required to meet the objective within the time. From this the team can establish the extent to which such resources need to be supplemented to satisfy the objective in time, which in turn will materially influence the method of contracting and the allocation of responsibilities.

2  The form of the financial resources which are available to the employer. These will consist of one or more of the following:

   ● cash

- commercial loans
- credit provided by the foreign firm or firms undertaking the work usually supported by a guarantee from their export credit agency (supplier credit)
- credit provided by a bank in the country of the exporter to the employer to enable him to pay the exporter supported by a guarantee from the export credit agency of the country concerned
- a loan from an international lending agency
- project or 'off-balance sheet' finance which is provided by banks against the security of the profits expected to be generated by the project itself
- non-recourse finance, i.e. forfeiting
- barter or counter-trade.

A general description of each of the above methods of financing is contained in the author's book *The Art of Tendering* (Gower, 1986); more specialized information is contained in the Euromoney publications *Trade Financing* (2nd edn, ed. Charles J. Gmur, 1986) and *Project Financing* (5th edn 1988, 6th edn to be published in 1995). If payment is to be made otherwise than by cash, the selection of the most appropriate method or combination of methods of financing is an important part of the planning process. What must be recognized is that each method may impose some restraint on the employer's freedom of action in respect of how he contracts, from which country he purchases the goods/services concerned, the terms upon which he does contract or possibly all three together. These constraints and the consequences which follow from them need to be considered alongside the purely financial considerations, and reference will be made to them in the following chapters.

3  The money available and the profitability level required, and how these relate to the capacity and to all other major design factors affecting cost and, again, time. Using the term 'capacity' to cover broadly both size and other design criteria, then time, price, and capacity have largely a fixed relationship. If one has a certain value, so do the other two;

alter one and you alter at least one of the others. This may be described as the second law of contracting. If, for example, it is once established that the logic of a situation is that the capacity required cannot be met within the price limit set by management, or only if the time is extended, then management must be informed at the earliest possible moment so that they have the opportunity to reconsider and, as necessary, redefine the objective. It is no use hoping that somehow the price will come out all right on the day or that savings in time can be achieved by shutting one's eyes to reality. It just does not work that way.

# Method and responsibility

From the definition of the objective in time and a study of the resources both available and required the team can proceed to the planning of the method to be used and the responsibilities to be allocated to achieve the objective. It is never sufficient to say that certain goods are to be supplied, plant manufactured, or works constructed by a defined date without at the same time thinking of what might be called 'the three Ws'. This then is the third law of contracting: 'that for each contract/project there must be stated: what – by whom – and by when'.

Supposing for instance that the planning team was concerned with the installation of a new machine tool in a factory. The steps to be taken, by whom and by when, might be set down in tabular form, as shown in Table 1.2. It will be seen that not all these steps are consecutive; thus preparations to receive the machine can proceed concurrently with its manufacture. It may be found on completion of such a table that the total time is greater than that which management have allocated. If the objective is to be achieved within that time then some of the above steps may have to be modified; perhaps a single tender negotiated to save time in tendering. This may of course in its turn affect the overall cost, thus demonstrating once again the essential inter-relationship between the various steps in the planning process.

In deciding upon the allocation of responsibilities under the

**Table 1.2   Example of activities and responsibilities of planning team**

| ACTIVITY | RESPONSIBILITY | TIME | |
|---|---|---|---|
| | | START | FINISH |
| | | Week number | |
| Management and co-ordination | Works manager | 1 | 56 |
| Preparation of performance specification | Works engineer | 1 | 4 |
| Prepare and issue inquiries to vendors | Buying office | 4 | 5 |
| Preparation and submission of tenders | Manufacturers | 5 | 11 |
| Receipt of quotations and their evaluation | Buyer in conjunction with works engineer | 12 | 14 |
| Authority to order | Works manager | 14 | 15 |
| Placing of order | Buyer | 15 | 16 |
| Detailed design | Contractor | 16 | 20 |
| Manufacture | Contractor | 18 | 50 |
| Preparations to receive and install machine | Works engineer | 38 | 50 |
| Delivery | Contractor | 50 | 51 |
| Receipt of delivery, unloading, and installation | Works engineer under contractor's supervision | 50 | 52 |
| Commissioning | Contractor | 52 | 56 |

above sort of table there are a number of possibilities to be considered, remembering always that several of the decisions are interdependent and that a decision on any one will almost certainly limit the power of choice on at least one other. The various allocations are considered in detail in the next chapter.

# 2 The contract plan

At the very outset a contracting plan needs to be prepared for the total project, not just for the letting of the principal contracts, but for every activity which has to be carried out to bring the project to its conclusion, including those which are to be performed by the employer himself. Nor in its totality is it concerned solely with engineering and construction. It should cover the provision of funding and all those associated activities such as purchase of land, obtaining of wayleaves, planning permissions and the like and even recruitment of staff/labour and agreements with the unions for working at new locations or with different operating procedures. With a new process plant or other production facility it may need to cover the conclusion of offtake agreements with future purchasers of the product since these may be a vital part of the financing arrangements for the construction works. Indeed with a project which is to be financed primarily on the security of the profits to be expected from its operation, such as a new gas-field, the Channel Tunnel, or in certain countries new motorway construction, the lenders will be concerned with ensuring that every item which can possibly affect the level of profitability has been taken into account in the planning process.

The same approach should be adopted by any employer concerned with a new project, large or small, since too many projects have failed to produce their intended benefits because of a failure to anticipate, plan for and implement those associated activities.

Having drawn attention to that issue it is intended within the scope of this work to concentrate on just those actions which are related directly to engineering and construction works. Through its allocation of responsibilities, it is the contracting plan for these which establishes the basis of the contracts to be placed and has a material effect upon the project cost and programme.

There are two steps involved in the preparation of the plan. First the identification of everything which has to be done in order to bring the project to fruition. Secondly the division of those tasks between the employer and all other parties involved; consultants who may be engaged to act on the employer's behalf, contractors providing work and services and equipment or material suppliers. Table 2.1 sets out the full list of tasks which may be involved and from which the employer can select those which are applicable in any given case. Tables 2.2, 2.3, 2.4 and 2.5 describe the principal ways in which these tasks are typically divided between the parties involved. The division by task is also a division by responsibility and Tables 2.2 to 2.5 are given in ascending order of the level of responsibility accepted by the employer beginning with that in which he accepts the minimum.

**Table 2.1   Table of activities**

DESIGN
System, process or conceptual
Equipment/piping/electrical layout
Civil/building, performance, outline or functional
Equipment/piping/electrical detail
Civil/building detail
Temporary works

EQUIPMENT
*Supply*
Manufacture and pack
Delivery FOB
Ship

Deliver port to site
Spares
*Installation*
Provision of labour, unskilled and tradesmen
Supply of equipment and/or facilities necessary for installation
Supply of equipment and/or facilities necessary for no-load testing

CIVIL AND BUILDING CONSTRUCTION
Provision of labour, unskilled and tradesmen
Provision of supervision
Supply of materials
Supply of construction plant and facilities

COMMISSIONING AND ON-LOAD TESTING
Supply of operating labour
Operating supervision
Maintenance staff
Equipment and facilities necessary for the carrying out of the tests including process materials

SOFTWARE
Supply of system and equipment handbooks and manuals including if necessary their translation
Supply of spares lists
Supply of 'as-built' drawings
Supply of computer programs

TRAINING
Provision of instructors for training at works and on site
Provision of training facilities and aids

OPERATION
Provision of managerial and technical staff necessary for initial operation

MANAGEMENT AND CONTROL
*Overall project management as between:*
The employer and outside agencies and government departments
The employer and consultants engaged on the design of the project
The employer and the main contractor
The employer and others working directly to the employer, e.g. suppliers of free issue items, and the main contractor
The several contractors and suppliers each contractually responsible directly to the employer
Contract administration including certification of payments
Quality assurance
Industrial relations

The division of responsibilities for the carrying out of the above activities will be described under the following four headings:

1   Full turnkey contracting
2   Partial turnkey contracting
3   Traditional client co-ordinated contracting
4   Management contracting

It is recognized that the headings given to Tables 2.2, 2.3, 2.4 and 2.5 are not terms of art, and that within each there are, in practice, variations in the manner in which the tasks and therefore the responsibilities are divided. Nevertheless they do broadly represent the four alternative methods generally in use today for providing the framework within which the contractual relationships and responsibilities of the parties can be formulated.

After the analysis of the four methods the next section will discuss their respective advantages and disadvantages and propose a set of criteria to guide the employer in making his choice as to which to adopt.

It is recognized that in addition to the four methods referred to above there are also projects today which are being carried out on the basis that the contractor designs and builds the project as well as finances and operates it. It is outside the scope of the present work to discuss in any detail the method of contracting for such projects and therefore a brief outline only is given at the end of this chapter.

## Full turnkey

The term 'turnkey' is used here in its original sense of being a contract under which 'the driller of an oil well undertakes to furnish everything and does all work required to complete the well, place it on production and turn it over, ready to "turn the key" and start the oil running into the tanks' (The United States Second Circuit Court of Appeals in *Retsal Drilling Co. v Commissioner of Internal Revenue* 127 F 2d 355 at 357). Duncan

Wallace in his book *Construction Contracts* says '*Principles and Policies in Tort and Contract* (Sweet and Maxwell, 1986) has suggested that the expression "turnkey" should only be used if the design responsibility is that of the contractor. While it is agreed that under a turnkey contract, design always *is* the responsibility of the contractor, the distinguishing feature is not just that but in the all-embracing nature of the contractor's obligations, otherwise it would follow that every E & M contract let under a standard form such as Model Form MF/1 of the IEME would be "turnkey" which is plainly not the case'. It means therefore that there is a single contractor who undertakes the entire responsibility for the project from the design through construction and commissioning to the handing over of the project to the employer who has only to 'turn the key' (see Table 2.2). In practice it is not quite so simple as that since there are some obligations which, necessarily, the employer has to perform such as stating the requirements which the project is intended to fulfil and making available the site. But as regards those relating directly to the execution of the project in all its aspects these are to be borne by the turnkey contractor. It also follows from this that the employer needs to place on the turnkey contractor far more onerous obligations, e.g. relating to guarantees for performance and defects liability, than are to be found under normal standard conditions of contract for plant works (see further p. 37).

There are two other terms which are to some degree synonymous with 'turnkey'. These are 'package deal' and 'design and build' and there are variants on these. There is no standard industry form for a package deal contract, which is the term generally used in the building industry for the type of contract approximating to turnkey. In particular it is important for any employer offered such a contract form by a contractor to ensure that it really does place on the contractor the full responsibility of the design being fit for purpose.

As regards 'design and build' the current standard industry forms produced by the Joint Contracts Tribunal and the Institution of Civil Engineers appear only to place on the contractor the responsibility for his design being carried out

## Table 2.2  Full turnkey responsibilities

| EMPLOYER / CONSULTANT ENGAGED BY EMPLOYER | TURNKEY CONTRACTOR | SUB-CONTRACTORS AND EQUIPMENT SUPPLIERS |
|---|---|---|
| | *Performance and design* | |
| States required performance and quality. | Undertakes total design responsibility to the employer. | Undertakes such design responsibility to the turnkey contractor as stipulated in the sub-contract or purchase order. |
| | *Equipment supply* | |
| The obtaining of any necessary import licence. Obtaining of duty-free import status if appropriate. | Undertakes total responsibility to the employer for the suitability, performance quality and delivery. | Undertakes responsibility to the turnkey contractor according to the terms of his sub-contract or order. |
| | *Equipment installation* | |
| Site availability. Assistance with visas and work permits. | Undertakes total responsibility to the employer for the performance of the work. | May undertake to the turnkey contractor either no responsibility, or technical supervision of installation, or full supervision of installation, or full responsibility. |
| | *Commissioning* | |
| Material supply, if any required for on-load testing. Infra-structural services, e.g. incoming gas/electricity. Provision of operating personnel. Assistance with visas and work permits. | Undertakes total responsibility to the employer for the work other than that to be done by the employer. | Usually limited to the provision of commissioning engineers for the items within their scope of supply. |
| | *Software and spares* | |
| Defines system requirements: language, maintenance system, extent of spares holding wanted, stock control and re-ordering system to be used. | Undertakes total responsibility to meet employer's requirements. | Provides for their scope of supply: equipment handbooks, spares lists, as-built drawings, computer operating programs. |
| | *Civil and building construction* | |
| Site availability. Assistance with visas and work permits. Obtaining any necessary | Undertakes total responsibility to the employer for the works. He may either carry these | Undertakes responsibility to the turnkey contractor according to the terms of his sub-contract. |

| | | |
|---|---|---|
| planning permissions or licences. | out directly or sub-contract as he chooses. | His responsibility will be reduced to the extent that the turnkey contractor: nominates particular suppliers or sub-sub-contractors, or provides common services, e.g. concrete batching plants, camps and messing, quarries. |
| | *Training* | |
| Provision of the required number of trainees with needed level of education and language fluency. | Setting up and carrying out of training programmes as needed to enable employer to operate and maintain the project. | Responsible for in-works training for their scope of supply. |
| | *Initial operation* | |
| Operation of the project once it has been taken over, subject to any requirement for the contractor to re-test at the end of the defects liability period. | Provision of technical and managerial staff as may be required. Re-testing of plant, if required, at the end of the defects liability period. | Provision of technical staff to the turnkey contractor as he may require. |
| | *Management and control* | |
| Management of the interface between himself and outside agencies providing services or supplies to the project or granting licences. Monitoring of the turnkey contractor's performance. | Total management and control of the project and of all designers, contractors and others responsible to him. Management of the inter-face between himself and the employer as regards the activities for which the employer is responsible. | Management of the sub-contractors and suppliers responsible to him. |

with reasonable skill and care. They are based on the design being sub-contracted either to a separate designer, e.g. an architect, or to a consulting engineer employed by the contractor or perhaps to the contractor's own in-house design organization. Unlike the normal form of turnkey contract as developed in the plant contracting industry, there is assumed to have been considerable design carried out by the employer pre-contract which the contractor is then required to take over. Indeed he may even be required to take over the employer's pre-contract design team. This can raise difficult questions as to design responsibility if the contractor's liability is not limited to negligence (see p. 58).

Neither of these two types of contract form are truly 'turnkey' in that they do not unequivocally place on the contractor full responsibility for fitness for purpose, or the liabilities which should be associated with that responsibility, nor do they limit the role of the employer in a way which would be consistent with the turnkey concept.

Nearer to the turnkey concept are certain individual forms of contract which have been developed by the Department of Transport and the Scottish Department of Roads which do place upon the contractor the complete design responsibility.

## Partial turnkey

With any form of partial turnkey contracting the division of work and responsibilities as between the employer, his consultants and the turnkey contractor is necessarily less clear-cut and subject to variations to suit the wishes of the parties. To the extent that the employer now undertakes certain work either directly or through consultants or other contractors independently of the turnkey contractor, the employer will increase his level of responsibility, both for the work itself and the co-ordination of that work with that for which the turnkey contractor remains responsible. Perhaps the most common form of arrangement is that in which the turnkey contractor undertakes responsibility for work within what is often referred to as 'battery limits', i.e. the main process or production plant itself, whilst the employer contracts separately for the supporting facilities. The employer may also wish to have a close involvement in the design of, and supply of equipment for, the production plant. But in so doing he must balance whatever advantage he believes he gains, against the resultant diminution in the turnkey contractor's contractual responsibilities. What he cannot do – although many make the attempt – is to dictate to the turnkey contractor how he should perform the work, whilst seeking to hold him wholly responsible for the results. Here I do agree with Duncan Wallace's (*op. cit.*) comments on the abuse of turnkey contractors. In my view the only sensible division of activities, and therefore of responsibilities, as between the employer and the turnkey contractor is that the employer limits his involvement to those activities which do not impact directly on the production plant. For example, he could contract separately for the landscaping, the perimeter fencing and lighting, the office block and the gatehouse.

Table 2.3 is based on the concept that authority and responsibility go together so that the contractual liability is accepted by the party undertaking the work or having the right to give instruction as to how it is to be performed.

**Table 2.3   Partial turnkey responsibilities**

| EMPLOYER/CONSULTANT | TURNKEY CONTRACTOR | OTHER CONTRACTORS AND EQUIPMENT SUPPLIERS |
|---|---|---|
| *Performance and design* | | |
| Responsible for that which he designs, e.g. off-site facilities. Will state required performance and quality of that to be designed by the turnkey contractor. May require to approve turnkey contractor's drawings but should not interfere with turnkey contractor's discretion otherwise may be held to assume design responsibility. | Undertakes total design responsibility for his scope of work subject to no interference by the employer. | Both sub-contractors and suppliers to the turnkey contractor and firms working direct to the employer, will be responsible according to the terms of their respective contracts. In practice contractors who work directly to the employer on the provision of off-site or ancillary facilities will not usually be responsible for design. |
| *Equipment supply* | | |
| For equipment within the plant for which the contractor has design responsibility as for total turnkey. | As for total turnkey. | As for total turnkey for contractors/suppliers to turnkey contractor. For other contractors/suppliers working for the employer |

|  |  | their responsibility will be according to the terms of their respective contracts. |
|---|---|---|
| *Equipment installation* | | |
| Usually as for total turnkey, except for those items which the employer elects to contract for himself. His responsibilities will then be according to the terms of the respective contracts which he places. | As for total turnkey for the equipment included in his contract. | As for total turnkey. |
| *Equipment commissioning* | | |
| As for total turnkey except for items which the employer has decided to contract for separately. | As for total turnkey except for items which the employer has decided to contract for separately. | As for total turnkey for contractors/suppliers to the turnkey contractor. For other contractors/ suppliers working direct to the employer their respon- sibilities will be according to their respective contracts. |
| *Software and spares* | | |
| Normally as for total turnkey but with the same qualifications as for equipment supply. | Normally as for total turnkey but with the same limitations as for equipment supply. | Normally as for total turnkey but with the same limitations as for equipment supply. |
| *Civil and building construction* | | |
| As for total turnkey but in ad- dition takes responsibility for work outside turnkey contractor's scope. | As for total turnkey for work within his scope. | As for total turnkey. |
| *Training* | | |
| As for total turnkey. | As for total turnkey. | As for total turnkey. |
| *Initial operation* | | |
| As for total turnkey. | As for total turnkey. | As for total turnkey. |
| *Management and control* | | |
| In addition to the responsi- bilities as for total turnkey the employer will be responsible for the management and co- ordination of work as between his own activities and those of the turnkey contractor. | Responsibility for the management and control of the project as defined within the scope of his activities. | As for total turnkey. |

# Traditional client co-ordinated

As its name implies this is the traditional form of contracting developed in the UK primarily within the building and civil engineering industries, and extended to multi-disciplinary projects which include large-scale electrical and mechanical engineering works (see Table 2.4). This method is based on the principle that the employer is responsible for the design of the permanent works, the supervision of their construction and, if there is more than one main contract, for the management and co-ordination of the separate contracts into which the project is divided. For projects of any size or complexity few client organizations across the world possess the engineering and project management resources to tackle these tasks and so over the years the professions of architecture and consulting engineering, and more recently quantity surveying, developed to provide these services. At the same time the division of the functional responsibilities of contractor and consultant became formalized in the forms of contract and practices of the professional institutions, especially within the UK.

At one time therefore this method came to be regarded as the norm. It still is widely used, particularly within the public sector,

**Table 2.4  Traditional client co-ordinated responsibilities**

| EMPLOYER/CONSULTANT | CONTRACTORS AND MAIN EQUIPMENT SUPPLIERS | SUB-CONTRACTORS/ SUPPLIERS |
|---|---|---|
| | *Performance and design* | |
| Responsible for the overall project design, the design of the buildings and civil engineering works (other than the detailed design of specialist items to be supplied and installed by nominated sub-contractors), the functional design of plant and equipment. | Each contractor will be responsible within the terms of his own contract. In general this will mean: buildings – none, civils – temporary works and nominated sub-contractors if so specified, plant – design to meet requirements specified by the employer. | Each sub-contractor and supplier will be responsible to the main contractor employing him within the terms of his sub-contract. |

### Equipment supply

May nominate certain specialist suppliers to the building and civil contractors. May supply some free-issue items.
Will place separate orders with individual suppliers and so through his consultant must now accept the responsibility for the inter-face between them and the compatibility of the equipment supplies.

For building and civil contractors the position is the same as for partial turnkey. For the individual equipment suppliers they will each now be responsible to the employer within the terms of their respective contracts for their scope of supply subject to the due performance of their obligations by others to the extent these inter-relate.

Sub-contractors and sub-suppliers will be responsible to those placing orders with them for the performance of their obligations within the scope of their supply subject to the same proviso as for the main contractor/supplier.

### Equipment installation

The employer must decide on the extent to which he wishes to be involved in the co-ordination of matters relating to the employment of site labour by the various contractors who will each be working to him.
Alternatively the employer may appoint one installation contractor for the whole site or provide the installation labour himself. In addition as for total turnkey.

Each contractor/supplier will be responsible within the terms of his contract. This may vary as between: no responsibility if the employer appoints a separate installation contractor, or responsible for technical supervision only or full supervision, if the employer is providing the installation labour for the distinction, or full responsibility.

His responsibility will be no greater than that of the main supplier or contractor to whom he is responsible. It could be less if the main supplier/contractor is himself fully responsible but elects either to install himself or to require the sub-supplier/contractor only to provide supervision.

### Equipment commissioning

As for partial turnkey but with the addition of the responsibility of the commissioning of the project as a whole.

Each main contractor/supplier will undertake the commissioning of his own section.

Each sub-contractor/supplier will be responsible to his main contractor as per the terms of his sub-contract.

### Software and spares

The employer through his consultant must co-ordinate the work of the provision of spares and software in accordance with his requirements. The adequacy and sufficiency of the whole to meet the project's operating requirements is now his responsibility.

Each main contractor/equipment supplier will be responsible for meeting the employer's requirements in respect of his section of the project.

Each sub-contractor/supplier will be responsible for meeting his main supplier/contractor's requirements for his scope of supply.

### Civil and building construction

| | | |
|---|---|---|
| Much the same as for partial will turnkey but the further he sub-divides the work into separate contracts the more responsibility he takes. | Each contractor will be responsible for his own section of the project subject to the proper performance of work by others for which the employer will be responsible to him. | Each sub-contractor be responsible to his main contractor within the terms of his sub-contract. |

### Training

| | | |
|---|---|---|
| The employer will be responsible for the establishment of the overall training and for the co-ordination of the training activities. In addition he will have the same responsibilities as for total turnkey. | Each main contractor/supplier will be responsible for such training as is specified in his contract for the section of the project for which he is responsible. | Each sub-supplier/contractor will provide training for his own equipment as specified. |

### Initial operation

| | | |
|---|---|---|
| Operation of the project once it has been taken over. | Provision of technical staff as required for the section of the project for which the individual contractor/supplier is responsible. | Provision of technical staff to his main supplier/contractor as required. |

### Management and co-ordination

| | | |
|---|---|---|
| The employer will be responsible for the overall management and co-ordination of the entire project. This will cover the work to be done by the consultants and the main contractors and suppliers directly responsible to him together with the supplies and services to be provided by his own departments or outside agencies. The detailed co-ordination of the work of the main suppliers and contractors will be handled for the employer by his consultants who will also be responsible for contract administration. | Responsible for the management and control of work to be performed under his individual contract. | As for the main supplier/contractor. |

both in the UK and countries overseas who are members of the Commonwealth and whose public service traditions are still founded on those of the UK. It is the method preferred by the international lending agencies. The reasons in both instances are basically the same, namely:

1   It places the design and the project management responsibility in the hands of 'professional' architects, engineers and quantity surveyors, whom it is assumed will act solely in the client's best interests without regard to their own commercial benefit.
2   It gives the client the benefits to be gained from the maximum competitive tendering for each individual section of the project. For example on a simple process plant where the civil engineering work may comprise no more than 10 per cent of the total costs, letting the civils as a separate contract from the plant works will avoid the payment of any turnkey contractor fees and ensure that competitive tenders are obtained for both. Further – and this is a point of concern to governments in the developing countries and to the international lending agencies – it will allow the civil works to be tendered for by local contractors. It is a method which satisfies well the concept of 'public accountability'.

There are of course disadvantages – which is why other methods have been developed and are favoured for certain types of project or contracting situations. This issue is discussed fully in the next section.

## Management contracting

The management of a project, both as a whole and its component activities, such as design and construction, has long been recognized in the US as a separate discipline, and this concept has now become widely accepted within the UK. The issue is then how the project should be managed for the benefit of the employer and three differing approaches can be distinguished:

1   Project management. The employer appoints a professional project manager to act on his behalf in the management of the project.
2   Construction management. Under this form the construction manager enters into a direct contract with the employer for the management of the construction of the project and may undertake a responsibility in relation to time and cost. All other consultants and contractors also enter into direct contracts with the employer.
3   Management contracting. Generally under this form the employer appoints one contractor who carries out none of the work himself but sub-contracts all of it to works contractors responsible directly to himself but under the control of the employer, through his project manager. The design and other consultants are appointed by, and responsible to, the employer.

The appointment of professional project managers has become much more widespread in UK practice and is specifically provided for in the new engineering contract (see p. 191). The main problem with such appointments lies in the degree of responsibility which the project manager owes to the employer and possibly also to the contractors. This issue is discussed further later (see p. 30). Their contract does not affect the contractual relationship between the employer and others and so will not be discussed further.

Construction management in its usual form does, however, affect the employer's contractual relationships with others. The employer is placed in direct contract with the various trades contractors who may well include some whom under the traditional client co-ordinated method would have been nominated sub-contractors to the main contractor. The employer also being in direct contract with the other professionals, such as the architect and structural engineer, may find himself faced with significant tasks of co-ordination and administration which may necessitate the appointment additionally of a project manager unless his contract with the construction manager is extended to encompass those tasks. This is quite contrary to the original concept of construction management. It was the construction manager who was supposed to manage both design and construction and be responsible for the design programme, monitoring the design progress and for the buildability of the design. In the US where the concept originated the construction manager is the leader of the team both for the management of the design and for construction. This is not the usual position in the UK where the leader appears to be the employer.

Two other issues arise. First, that of the liability of the construction manager for the work of the various trades contractors. It can be argued that the construction manager should have a liability for them, since otherwise the employer, by having a multitude of separate contractors each working to him and each likely to blame the others if anything goes wrong, would be left in practice without an effective remedy. (Elizabeth Jones in the *International Construction Law Review* 1993, at p. 353, argues this way.) Against this it is suggested that making the construction manager responsible for the trades contractors removes him from being a part of the employer's team and re-creates the climate of adversarialism a reduction in which it was intended that this method of contracting should achieve.

The second issue is that of the liability of the construction manager himself. He will clearly be responsible to the employer for exercising reasonable skill and care in the performance of his duties and may, depending on the definition of his scope of responsibility, be under a greater duty (see p. 58).

Further it is considered that the contractual duty of the construction manager to the employer to supervise the work of construction or installation would include the responsibility of being familiar with any particular methods of work to be employed and knowledge of any manufacturer's instructions to be applied. In this respect and depending on the terms of the particular contract it seems that the construction manager's responsibilities for supervision could be greater than those of an architect or consulting engineer.

The third method, management contracting, has lost something of its one-time appeal. Under this method it is normal for the management contractor to be responsible to the employer for the work of the works contractors with whom he is now in direct contractual relationship, but ultimately his liability for a breach of contract by a works contractor is generally limited to the amounts which he is able to recover from that works contractor in arbitration/litigation. (For the possible effects of these provisions, see p. 292.) In the absence of such a limitation his liability would hardly be different from that of a normal main contractor. That his liability should extend to being fully responsible for failures in time, price or standards of work of his sub-contractors is a view which has often been expressed by traditionally minded quantity surveyors. Such a view retains the time-honoured adversarial relationship and with it the role of the professional quantity surveyor acting for his client in opposition to the contractor, and negates the very purpose of the managing contracting system.

The difficulty with management contracting is that it does not place the management contractor firmly on either the employer's or the contractor's side of the table and bitter experience has taught the author that you cannot sit on both. The greater the degree of responsibility which the employer seeks to place on the management contractor in terms of completion to time and to a predetermined cost, the more closely his role resembles that of a conventional main contractor and the more strongly is re-created the adversarial contractual relationship between employer and contractor which it was one of the objectives of the management contracting system to remove. Again the management contractor under a standard form such as that produced by the JCT, although required to co-operate with the employer's professional team responsible for the design, is not himself responsible for the management of the design process. This is clearly a great weakness in that it dilutes his responsibility for the programme.

The respective responsibilities of the project manager, construction manager and management contractor as they are commonly found in contracts in current use are illustrated in the charts in Figure 2.1 but it must be remembered that this is an area in which standard forms play little part and most contracts

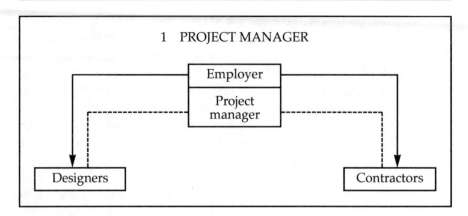

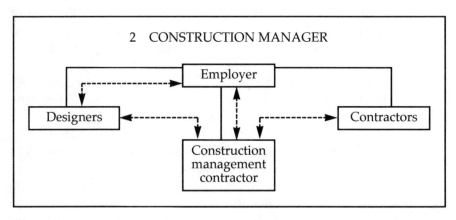

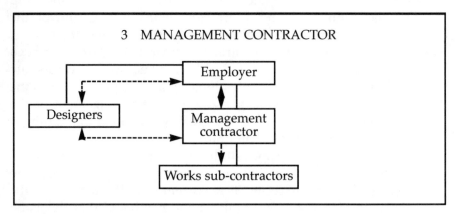

**Figure 2.1  Management contracting forms of responsibility**

are developed by individual clients or contractors. In practice therefore the responsibilities may vary from one contract to another.

Table 2.5 is based on the role of the construction manager being limited to management, with the 'hands on' construction work being undertaken by trades contractors working directly to the employer who remains through his project management team responsible for overall co-ordination. This method appears today to be that most commonly used in the UK.

**Table 2.5   Construction management responsibilities**

| EMPLOYER | CONSTRUCTION MANAGER | CONTRACTOR / SUPPLIER |
|---|---|---|
| | *Performance and design* | |
| Defines project. Appoints consultants responsible for the design. Overall monitoring. | Co-operates with employer's design team. | Carries out specialist equipment design. Civil contractors design their own temporary works. |
| | *Equipment supply* | |
| As for total turnkey but in addition places contracts on the recommendation of the construction manager. | Prepares bid lists, invites and analyses bids and prepares bid recommendation and contract documentation for employer approval and signature. Carries out quality control and expediting. | Undertakes responsibility to the employer according to the terms of his order but acts in accordance with instructions received from the construction manager. |
| | *Equipment installation* | |
| As for total turnkey but in addition places any separate installation contracts on the recommendation of the construction manager. | Decides on the allocation of work and whether it is to be undertaken by the equipment suppliers, or under their supervision with a separate installation contract. May provide common services, e.g. accommodation and messing, shipping to the site. Manages the performance of the work. | Undertakes responsibility to the employer either to install or provide installation supervision and acts in accordance with instructions of the construction manager. |
| | *Commissioning* | |
| As for total turnkey. | Manages the work to be carried out by others. | Undertakes the commissioning of the |

|  |  | plant included within the scope of his supply. |
|---|---|---|
| *Software and spares* | | |
| As for total turnkey. | Manages the provision by others of the spares and manuals etc. to meet the employer's requirements. | As for total turnkey in accordance with instructions received from the construction manager. |
| *Civil and building construction* | | |
| As for total turnkey and in addition places contracts on the recommendation of the construction manager. | Decides on the division of the work into work packages. Prepares bid documentation and bid lists, invites and analyses bids, submits recommendation and contract documentation to employer for his approval and signature. Carries out contract administration, including progress and quality control. May be authorized to issue payment certificates but this is more likely to be undertaken by the project manager. | Undertakes work in accordance with the terms of his contract and instructions received from the construction manager. |
| *Training* | | |
| As for total turnkey. | Prepares overall training programme and manages the carrying out of training by others. | Carries out training in accordance with the terms of his respective contract and instructions of the construction manager. |
| *Initial operation* | | |
| As for total turnkey. | Will arrange for the provision of any managerial staff which employer may require. | May provide technical staff to assist the employer as instructed by the construction manager. |
| *Management and control* | | |
| Overall management of the project through his project manager and management of the interface between himself and outside agencies as for total turnkey. | Management of the trades contractors responsible for the procurement, installation, construction, commissioning and training, control of quality, programming of construction and contract administration under the direction of the project manager. | Management of the sub-contractors and sub-suppliers responsible to him. |

# Advantages and disadvantages of different methods of contracting

## Selection of criteria

The sum of the risks and responsibilities involved in the execution of the planned project to not change because of the method of contracting which is adopted. They are a function of the nature of the project itself and its location related to the technology to be employed and the physical and political conditions under which the work is to be executed. What the particular method of contracting chosen will do is to allocate the risks as between the parties involved and in so doing affect the likely outcome of the project in terms of cost, time and performance.

Considering the four methods which have been discussed the advantages and disadvantages of each are summarized as follows:

### 1 Total turnkey

#### Advantages

1   Places maximum responsibility for the project in the hands of one organization and minimizes the need for the employer to employ his own resources or engage consultants. It has been the experience of the Department of Transport that the use of design and build contracts for roads has substantially reduced the staff on site, especially from the consultants, with resultant economies in cost. (See the paper given by Tony Holland of the Department at the Conference on the ICE Conditions of Contract Design and Construct organized by IBC Legal Studies and Services Ltd held on 7 December 1992.)

2   It should bring about the completion of the project within the shortest possible period of time.

3   By making the design part of the competitive tender it encourages innovation and economies and should result in lower project costs.

4  It should enable economies of cost to be secured by the synchronization of design, procurement and construction so avoiding the delays and diseconomies inherent when designers, purchasing agencies and construction contractors belong to different organizations.

5  It should reduce to a minimum claims against the employer for extras since it is up to the turnkey contractor to deal with claims arising from the delay or bad performance of one sub-contractor on the work of another. This means that the out-turn costs should be very close to the original contract price. However these advantages will only be secured if the employer:

   ● Has selected the right turnkey contractor in the first instance and 'right' here is usually not the apparent cheapest. The technical, managerial and financial resources which the turnkey contractor possesses and is both able and willing to devote to the contract are of greater importance than the initial price.

   ● Was able at the time of tendering to define his requirements in sufficient detail to enable the turnkey contractor to give a firm price.

   ● After contract award does not make substantial and/or recurring changes in his requirements and leaves the turnkey contractor to get on with the work without interference either from his own staff or consultants. Of course the employer would be rightly concerned to see that the project is monitored to ensure that the work is being carried out in accordance with the contract terms, but he must not start trying to 'second guess' the contractor in terms of either design, procurement or construction. This is a temptation which it is often hard for either the employer's own engineers or consultants to resist.

*Disadvantages*

1  Once the selection of the turnkey contractor has been made

there is little opportunity for the employer to correct any mistake in the choice of firm concerned. Accordingly the contract must contain stringent guarantees and penalties and the employer must be satisfied that he has sufficient financial security from the turnkey contractor to enforce these should the need arise. Such guarantees must cover fitness for purpose, without the need for the employer to establish negligence, and run for a period long enough to establish that this requirement has been satisfied – a minimum of five years from completion.

2   Depending on the size and complexity of the project the employer may find that his choice of firms to compete for the work is very limited due to the increased costs of tendering and the scale of engineering, managerial and financial resources needed.

3   The contract price is bound to reflect the scale of the risks which the turnkey contractor is accepting, of the resources which he is required to employ and the relative lack of competition.

4   Against the advantages of the 'turnkey' form there is the undoubted risk that the contractor will be influenced in his decisions on detailed design, selection of vendors and construction methods primarily by commercial factors and that the eventual project, while meeting specification, will not incorporate factors of safety or of long-term life of the type upon which a professional consulting engineer would probably insist. In part the employer can guard against this in the preparation of the tender documents, analysis of the contractor's proposals and inspection of the work as it proceeds but as stated above once he has committed himself to the contract the employer is constrained in the extent to which he can require changes or insist on any modification of the manner in which the contractor intends to perform the work.

*Design and build.*   Although as indicated above design and build is not strictly a turnkey contract it does have certain of the advantages for the employer which turnkey contracting possesses. It should reduce the time for completion and produce economies in cost through the involvement of the contractor in the design and the inclusion of at least the detailed design within the competitive tendering process.

It also has some of the disadvantages such as a restriction on competition and the employer needs to take care to ensure that there is an independent check on the contractor's design and that he has the necessary contractual remedies in place.

### 2   *Partial turnkey*

*Advantages*

1   For the work which is the responsibility of the turnkey contractor then the same advantages apply as for total turnkey.

2   The employer is given the opportunity of contracting separately and probably more cheaply for the ancillary work which is out with the scope of the turnkey contract. This can allow him the chance to give work to smaller local firms.

*Disadvantages*

1   The employer must resist the temptation to undertake ancillary works which are necessary for the proper functioning of the works being undertaken by the turnkey contractor. If he fails to do this, or is prevented from so doing by local regulations or the method of financing, and the ancillary work is late, then the employer will have paid in the turnkey contract price for the speed of construction of that element but without achieving any overall economic advantages. A typical situation in which this occurs is where the turnkey contract is financed by export credits but the ancillary works have to be paid for out of the employer's own budget and either the money is not available when required or the bureaucratic procedures involved are such that contracts cannot be awarded at the right time.

### 3 *Traditional client co-ordinated*
#### *Advantages*

1 The employer obtains the benefit of independent professional design and supervision of the construction of the works.
2 Each work-package will be tendered for on a basis which will ensure the most competitive prices. The method is ideal if the work can be executed under a single main contract so that the co-ordination is limited to that between design and construction and between the contract work and any other associated activities.
3 The employer through his consultant or own engineering department retains control over the project and changes can be accommodated within the contractual procedures.
4 The employer is able to exercise choice over the firms asked to tender to whom work-packages are contracted, and can accordingly ensure that where he considers it appropriate preference is given to the employment of local firms.

#### *Disadvantages*

1 By entering into separate contracts for the equipment and the foundations the employer acts as a co-ordinator between two separate contractors. Information to and from each must pass through him which inevitably slows down the process of communication.
2 The contractor for the equipment, who must supply the necessary data for the foundation design, will not release this in final form until absolutely certain that it is correct. If released earlier, and changes had to be made, the employer could claim against him for any additional costs incurred due to the re-design.
3 Similarly the civil contractor will not act on preliminary or incomplete data but will wait until it is released in final form. Both have everything to gain, and nothing to lose, by waiting until they are clear to proceed without the risk of having to

make late changes. The plant contractor is unlikely to incur any penalty merely by being late in supplying data; the civil contractor will not only have a legitimate claim for extension of time for completion of his work, if he is delayed through the late receipt of design data, but may also have a claim against the employer for extension of preliminaries.

In contrast a main contractor who is solely responsible for the design of the plant and foundations and construction of the foundations is in a position to feed through information piece-meal, switch construction labour from one task to another and issue preliminary drawings to the field with 'hold' marked on them for specific items only. In doing this he takes certain risks, and may incur additional costs which he must now bear himself, but he will have allowed for this in his contingency. The more efficient and experienced as a main contractor he is, the less the contingency need be, and he now has every incentive both to minimize costs and to complete the project on the overall completion date, since he will have no claim either for extra time or costs because of his sub-contractor's default.

4   Any contract which the employer places himself must go through the administrative procedures peculiar to his organization and with which he is obliged to conform. These may involve, in the case of a public authority, inviting public tenders and obtaining formal authorization from a board before a contract can be placed. This inevitably takes time, whereas the main contractor responsible for the total project has the flexibility to gear his administrative and commercial actions to suit the requirements of the project programme.

## 4   *Management contracting*
### *Advantages*

1   Savings in time can be achieved in comparison with the client co-ordinated method without the employer having to commit himself to a turnkey contractor. This can be especially valuable where time is short and it is necessary to start construction on one work-package prior to the

completion of design on others and 'leap-frog' design and construction while handling the changes which this will necessarily involve – what is often referred to as 'fast-tracking'.

2 With construction management there can be savings in cost to the employer because of the 'hard-nosed' commercial attitude which the construction manager will bring to the engagement and control of the works contractors. This will be accentuated if the construction manager is on a bonus for bringing in the project under budget.

### *Disadvantages*

1 In the same way as in the client co-ordinated method the employer has to accept the risk of claims from one contractor by reason of the default of any of the others. His hope is that the management contractor will have acted to minimize the impact of these.

2 The employer will not know the out-turn cost of the project at the start, although he will expect that the budget from the management contractor should be reasonably accurate.

## Decision criteria

This section sets out the factors which can be relevant to the employer's decision as to which method of contracting to adopt.

### 1 Governmental regulations or guidelines
Government instructions may affect the decision on contract planning by:

● insisting on a firm or guaranteed maximum price. This would rule out the form of managing contracting described under method three. As referred to in the discussion on managing contracting (p. 31) by requiring the contractor to accept this obligation he becomes in effect a main or general contractor.

- in overseas developing countries, requiring that construction work be sub-contracted to genuine local contractors, i.e. firms either wholly or majority-owned by local shareholders. This could make the turnkey method impractical because the contractor would not accept this limitation as being compatible with the freedom of commercial and technical action necessary for him to accept turnkey responsibilities. The general issue of the use of local contractors and resources is discussed later on p. 122.

- a regulation requiring that the works be put out to competitive tender to the widest possible selection of contractors, with design being carried out by the authority's engineering department or consultants employed by them. Clearly this could only be met by the use of method three, client co-ordinated.

### 2  *Method of funding*

Unless the employer is providing the funds for the project out of his own budget – in which event he would be free to choose the method he considered the most appropriate – the lenders to the project will influence the method in any one of the following ways according to the method of funding selected:

- As indicated earlier if a loan is being provided by one of the international lending agencies, then normally the agency will require that the construction works and the supply and erection of related plant and equipment are contracted for separately – see para. 2.04 of the *Guidelines for Procurement under Asian Development Bank Loans.* Only exceptionally, where special processes are involved, will the agency allow the use of turnkey contracting.

- If the project is being financed by project financing, i.e. on the strength of the security of the project itself and the profits it is expected to generate, then the lenders are likely to prefer that it is contracted for on a turnkey basis since this can be expected to minimize the risk of additional funding being required and ensure that the project will be brought into commercial operation on the projected date.

- The provision of governmental aid, e.g. for the UK under the aid for trade scheme, the government will insist that the benefit of the contract(s) for which the aid is provided goes to contractor(s) of their own country. In practice this will favour turnkey contracting.
- Financing of the offshore portion of the project by export credits does not necessarily influence the method of contracting. It can be compatible with any of those discussed. However if the project is being wholly financed, say, by a combination of export credits and commercial loans then the lenders and credit agencies are likely to want the project contracted for either on a turnkey basis or with the use of a management contractor.

## 3 Project size, complexity and employer resources

The project size and/or complexity related to the managerial and technical resources in terms of both numbers, quality and relevant experience which the employer can himself make available, will have a major impact on the decision as to which method of contracting to adopt.

- If the project is basically simple and/or small-scale related to the employer's resources, then the preferred choice would generally be method three, traditional client co-ordinated, since it would be likely to produce the most economical results. This is because:

  - The employer avoids paying fees to a turnkey contractor for taking responsibility for the work of others. Depending upon the costing system employed by the firm acting as turnkey contractor, such fees may either include the firm's normal commercial overheads and profit or be limited to a handling charge, but they will invariably also include an allowance for risk, which will be higher the less familiar the main contractor is with the sub-contracted work.
  - The employer can invite competitive tenders for each separate contract from specialist firms of his own choice

and then select the most favourable offers received. The turnkey contractor, if he is a member of a larger group, may be required to place business with other group subsidiaries, but in any event is likely to be conservative in the placing of orders and to select firms who have served him well in the past. He knows that he cannot expect his suppliers fully to 'underwrite' his own liabilities to the employer and for his own protection will tend therefore to select firms with whom he has close commercial relationships and who can be expected to provide support to him beyond the strict terms of their contractual liabilities.

However, in particular cases the savings to be obtained by achieving an earlier completion date through the use of the turnkey method may outweigh these and show that this is the best overall solution (see further below under the heading, 'Time for completion').

● In the opposite case when the project is large and/or complex in relation to the employer's resources, then the choice should be either turnkey or the use of a managing contractor. However, in either case the employer must ensure that he has the necessary resources to enable him to make the right choice of contractor: if necessary by engaging consultants to prepare the project brief against which the selection is made and evaluate the proposals received.

### 4  *Time for completion*

With revenue-earning or cost-saving projects an assessment should always be made of the value of time which could be saved through the use of either the turnkey or managing contractor methods in comparison with client co-ordinated. This time-related saving should then be compared with the cost saving which would be achieved under the client co-ordinated method by the division of the project into separate work packages.

In certain instances, e.g. the need to have a facility operational by a specific date to catch a seasonal trade, the saving of time will become the overriding factor.

## 5  *Economy of design and operating costs*

One of the main counter-arguments to the proposition that the client co-ordinated method leads to lower capital costs is that there will be no element of design competition and any design produced by consultants or the employer's engineering department will tend to be conservative so that the project will be over-designed and therefore expensive. Further, any such design will not take into account potential manufacturing or construction economies which are the prerogative of the contractor. One possible answer is to allow the contractors when bidding to submit alternative bids based on their own design. However this has its own difficulties (see p. 132).

The validity of the argument is not easy to assess. In recent years consultants have become increasingly engaged by contractors for the provision of design services when the contractors have been bidding on a turnkey basis. They have also been exposed to commercial clients overseas who have demanded economies. The consultant's approach today therefore is generally more commercial than it was some years ago. However there remains nothing like competition for concentrating minds on the most economical solution and if the project is one which lends itself to design ingenuity then the potential for cost savings through including the design in the bid competition should additionally be taken into account when making the savings comparison referred to above under the heading 'Time for completion'.

A negative aspect of competition, however, can be its effect on operating and maintenance costs. Unless a contractor is specifically aware that the employer is both interested in, and has the capability to, take such costs into account in making his adjudication, then in his own interests the contractor must tender on the basis of the lowest possible capital costs. That will almost certainly mean that the employer will be faced in the future with higher maintenance and operating costs. Provided that the employer is allowed by his tendering regulations and the EC Procurement Directives to award the contract to a firm bidding other than the lowest initial costs then it is suggested that the employer should make it clear in his enquiry that operating and maintenance costs including spares over the expected life-time of the project will be assessed and taken into consideration when making the adjudication. This is specifically provided for in the regulations of the international lending agencies which state for example:

Bidding documents should specify the relevant factors in addition to price to be considered in bid evaluation and the manner in which they will be applied for the purpose of determining the lowest evaluated bid. Factors which may be taken into consideration include inter alia . . . operating costs . . . the availability of service and spares. . . . The factors other than price to be used for determining the lowest evaluated bid should, to the extent practicable, be expressed in monetary terms or given a relative weight in the evaluation provision of the bidding documents. It is also now provided in the EC Procurement Directives and the Utilities Directive that where the award is to be made to the most economically advantageous tender then criteria for evaluation must be stated either in the notice to be published in the official journal or in the invitation to tender, and where possible in descending order of importance.

*Guidelines for procurement under IBRD Loans,* issued by the World Bank

### 6  *Certainty of out-turn costs*
Provided that the employer has made up his mind in advance as to what it is that he wants and does not change it later on, then a lump sum turnkey contract is the best method of avoiding additional costs since it places the risk and responsibility firmly with the contractor and leaves the employer's architects and engineers with the minimum opportunity for initiating variations. With any other method, but especially managing contracting, changes are relatively easy to handle and costs have an alarming tendency to escalate over original budgets. 'Fast-tracking' will certainly save time but the price can be high and if the employer's budget is limited and additional funds would be difficult to raise, then he would be well advised to ensure that

the price is definitively settled in advance of construction, which of course means that the design must be as well.

There are managing contracting methods in which the price and design are refined in an iterative process of negotiations between the employer and the managing contractor and a maximum price established before construction starts, but it is difficult to see – other than perhaps quality – what advantage they offer over conventional methods. Certainly it cannot be time and if it is claimed that the employer gains in terms of cost from the collaboration between his architects and the managing contractor then equally he loses the price benefit of competitive bidding from main contractors.

## Summary and conclusions – conventional methods

1  The preparation of a contracting plan is an essential step in the execution of any project no matter how simple.
2  There is no single or perfect answer. Each plan represents a trade-off between conflicting interests; shorter time against lower capital cost; unified responsibility resting with the contractor against retention of control by the employer; design competition against Rolls Royce standards; employment of local resources against optimum cost/completion time.
3  The preparation of the plan, because of the trade-offs involved, cannot be the work of one department or function. Each must be represented on the planning team and management are only interested ultimately in the whole; they are not concerned with the bits or who does them. Unfortunately over the years the professions associated with construction would seem at times to have forgotten this, so that not only have activities been portioned out between different people when more properly they belonged together, but each portion has acquired merit for its individual worth and not necessarily for its contribution to the whole.
4  However much he may delegate to his consultants or turnkey contractor the ultimate responsibility for the project

always rests with the employer. It is essential therefore that he appoints at the commencement of the planning process and retains throughout the project an individual to act as his project manager who has the authority to represent him with all external organizations and to co-ordinate the activities of his own internal departments.

# Design, build, finance and operate

In recent years certain governments, including that of the UK, have wished to have particular projects implemented in the private rather than the public sector. This has been for two reasons. First, to supplement government funding with private finance. Second, to impose upon the project the commercial disciplines of the private sector which have been conceived as being stricter than those traditionally associated with the public sector.

The essence of such projects is that a group of private investors and contractors is formed which competes for a contract to construct the project, including a requirement to finance the construction and a concession to operate the project for a specified period – which is intended to be sufficient to allow the project to repay its costs and provide the investors with a reasonable return on their investment.

There are numerous problems associated with such projects, relating to their financing and providing the investors with a sufficient assurance of a return on their funds to make the investment a realistic attraction. However, our concern is with the method of contracting and the contractual relationships between the parties involved.

The first point is the nature of the contract between the government and the concessionaire. Is this in essence a concession, or is it a form of design and build contract to which have been added the obligations to finance and operate and the right to earn revenue, e.g. recover tolls?

The currently preferred form in the UK appears to be the latter, which was used by the Department of Transport for the

Queen Elizabeth II Bridge and the Second Severn Crossing.

The second point is the relationship between the concession company and the principal contractors involved in the construction of the project who are also shareholders in the concession company. The concession company needs the assurance that the project will be built on a competitively priced, turnkey basis with long-term guarantees on performance. The contractors/shareholders are looking for construction work and to be rid of both their contractual and investment obligations as early as possible and in any event on the expiry of a normal defects liability period.

The resolution of these conflicting objectives must be sorted out pre-concession contract stage. The further problem which may arise is the involvement of the concession company as operator in the design process. As commented earlier, a turnkey contract on a lump sum basis will not be successful if there are important design changes during the construction. It follows that the concession company's requirements relating to operation must be detailed before the lump sum price is agreed and as operators they must not subsequently make changes to their design requirements under the guise of design 'development'. If they do it is inevitable that there will be cost over-runs and disputes will occur.

# 3 Legal issues arising from the contract plan

The legal consequences which arise from the contract plan can be considered under four broad headings:

1 The establishment of the persons against whom the employer has (a) a right of action in contract and (b) who owe to the employer a duty of care.
2 The measure of damages which the employer may be able to recover against them either in contract or negligence.
3 The nature and extent of the liabilities which the employer may have to other persons either in contract or negligence and the measure of damages for which he may be liable.
4 The effect of pre-contractual discussions.

## Persons against whom the employer has a right of action; and who owe the employer a duty of care

### Main contractor/sub-contractor
The general rule of English law is that a contract only creates

rights and obligations enforceable by the contracting parties as against each other although as will be discussed later it may also have the effect of creating a duty on a third party not to interfere with the parties in their performance of the contract. This rule has given rise to a number of difficulties in sub-contracting, especially in relation to nominated sub-contractors and suppliers. The employer, having on the advice of his architect or engineer, selected a particular sub-contractor or supplier, is nevertheless not a party to the sub-contract between them. The contractual relationships as between employer–main contractor–sub-contractor may act so as to create a duty of care on the part of the sub-contractor towards the employer in negligence, or to provide the sub-contractor with a defence against a claim by the employer in negligence, but contractually the obligations of the sub-contractor are owed to the main contractor and not to the employer. By his action therefore in deciding to place a single main contract the employer has made his choice as to the party against whom he will have contractual rights.

However, in practice such rights may well prove to be illusory at the time when the employer wishes to enforce them since by then the main contractor may have gone out of business. Further, the obligations of the main contractor to the employer may not be such as to cover the issue in question. The general rule is clear that a main contractor is liable to the employer for the materials supplied and workmanship performed by a sub-contractor, whether nominated or not, unless such liability is expressly limited by the terms of the main contract itself. The principle behind the rule is that only through his contract with the main contractor can the employer have a contractual remedy for the deficiencies in the sub-contractor's work or materials and it is for the main contractor to protect himself in the warranties he obtains from the sub-contractor or supplier. However where the employer has taken it upon himself to investigate the suitability for his particular purposes of a specialist material – which under its trade name and from a specific supplier he then requires the contractor to incorporate into the works, without the contractor having any right to object – then the main contractor will not be liable if that material proves to be unsuitable for its purpose. As to whether or not the main contractor would be liable if the materials supplied were not of a merchantable quality would seem to depend on what limitations, if any, were imposed on the main contractor as to the extent of his ability to protect himself against the default of the nominated supplier. If not only the choice of supplier, but also the terms and conditions of supply were established by the employer, and these were restrictive of what otherwise would have been the main contractor's freedom of commercial action, then it could well be held that any liability on the main contractor for quality was excluded – see the House of Lords decision in *Gloucester County Council v Richardson* [1969] 1 AC 480.

The way in which the employer may be protected contractually in the above circumstances is if there is a collateral contract between himself and the supplier. Such a contract may be established expressly in the manner provided for in the *JCT 80 Building Contract* procedures by the architect obtaining from the nominated supplier the direct warranty under *Tender Form TNS/2* in favour of the employer. Alternatively where a supplier makes specific statements to a prospective purchaser about the quality and suitability of his goods, and in reliance on these statements the purchaser instructs the contractor to buy them, then a collateral contract may arise between the supplier and the purchaser. Should the goods then prove to be unsuitable the purchaser may be entitled to sue the supplier directly in contract. See *Shanklin Pier Ltd v Detel Products Ltd* [1951] 2 A11 ER 471, where the employer asked a paint manufacturer whether his paint was suitable below water level and in reliance on his statement that it was, specified it to the main contractor. In fact the paint was not suitable and it was held that the paint supplier in consideration of his product being specified had guaranteed its suitability for the job and was therefore liable under this collateral contract with the employer in damages for its breach. This case was cited with approval in *Greater London Council v Ryarsh Brick Co.* 4 CON LR 85, but in that case the evidence was such as to show that the GLC did not rely on any specific statements made by the supplier as to the suitability for the use of his particular bricks in the manner in which the GLC architect intended to use them in his design. As a result Ryarsh were held not to be liable to the GLC. The case illustrates the degree of precision and reliance which must be proved by the employer to exist in order for a claim on a collateral contract to succeed.

So far the discussion has been limited to the position of those involved in the construction operations as it arises in contract. At the time of writing the third edition of this book it was generally recognized, following the House of Lords decision in *Junior Books v Veitchi* 1983 AC 520, that under certain circumstances an employer could have a remedy in negligence against a nominated sub-contractor. Although that decision has not been formally over-ruled effectively, it can no longer be regarded as good law after the landmark decision of the House of Lords in *Murphy v Brentwood District Council* [1991] 1 AC 378.

In essence Murphy's case decided that as regards defective goods and buildings there was a clear distinction between liability in contract and liability in the tort of negligence. In contract a builder is liable to the employer, or the sub-contractor to the main contractor, for his defective work according to the terms of his contract. He is, however, only liable to a third party, e.g. a sub-contractor to the employer, in the tort of negligence for injury to persons or damage to other property of the employer. He is not liable to the employer for the defects in his work itself, no matter the seriousness of such defects. Defective work which causes the building to be worth less than it would be otherwise is classified as economic loss which is only exceptionally recoverable in tort.

The liability in negligence for injury to persons from defective work is reasonably clear. It will extend to cover those persons whom the builder should have had in contemplation as being likely to suffer injury if he does not take proper care in the performance of his work.

Liability for damage to 'other property' is more difficult. First, in this context what constitutes 'other property'? It seems clear that it would cover items such as computers which the employer has installed in the building under a separate contract and which are damaged, say by the fall of a defective ceiling constructed by a sub-contractor. However, consider the case of a boiler installed by a sub-contractor which explodes and damages other parts of the building. The cost of the replacement of the defective boiler itself is a loss recoverable only in contract and therefore only from the main contractor. If, therefore, the main contractor is not available to be sued, the employer, or his insurance company, will be left without a remedy. But in those circumstances can the employer recover at least the damage caused by the explosion of the boiler to the remainder of the building from the boiler manufacturer in tort as being damage to 'other property'? In Murphy's case it was suggested in judgements given by three of the Law Lords that he could do so provided he could prove that the explosion was indeed due to the negligence of the boiler manufacturer.

The difficulty with this approach, what is known as 'the complex structures theory', is how far it should be taken. For the purpose of defining 'damage to other property' the structure will normally be regarded as one unit. So defects in the work of a structural steel sub-contractor which weaken the frame of the building and cause damage to the floors or walls constructed by the main contractor or other sub-contractors will not, it seems, be regarded as causing damage to 'other property'.

The loss occasioned by all such defects is classified in law as 'economic loss', i.e. the building is simply worth less than it would have been had it been properly constructed, and economic loss is only exceptionally recoverable in tort.

What would be damage to other property has been much debated. It would appear that an electrical sub-contractor whose defective work positively malfunctioned and caused a fire which damaged other parts of the building could be held liable in negligence for such damage.

Effectively, however, the decisions in D & F Estates and *Murphy* have largely put an end to the expansion of the scope of the law of negligence as regards the ability of employers to claim damages from sub-contractors for the consequences of their defective work. If not formally over-ruled, the decision in *Junior Books* is no longer an authority on which any reliance can be placed. The trend noted in the previous edition of this book, to integrate contractual and tortious liabilities, has been completely reversed. The present intention of the House of Lords is clearly to maintain and even reinforce the distinction between the two.

The only possible exception to the rule, that an employer cannot bring an action in negligence against a sub-contractor for economic loss, would be if the employer could rely on a negligent misstatement by the sub-contractor under the rule established by the House of Lords in the case of *Hedley Byrne & Co. v Heller and Partners* 1963 and their later decision in *Caparo Industries plc v Dickman and Others* 1990.

In order to bring such an action the employer would have to show that there was 'a special relationship of proximity' between himself and the sub-contractor; that the sub-contractor knew that his advice was likely to be relied and acted upon by the employer without independent enquiry and it was so acted upon by the employer to his detriment. The factual basis upon which a nominated sub-contractor or supplier is appointed will not normally support such a finding. It is indeed more likely that, if the nominated sub-contractor in order to secure his nomination has made express representations about the quality or performance of the product that he is supplying, which might justify a *Hedley Byrne* liability in negligence, the same facts would support a claim in contract for breach of an implied collateral warranty on the principle of the *Shanklin Pier* case, which would be easier to establish. The obvious solution in practice is for the employer to obtain from a nominated sub-contractor or supplier an express collateral warranty (see further p. 288).

English law proceeds on the basis of a chain of contracts running from the employer to the main contractor, from the main contractor to a sub-contractor and on again to sub-sub-contractors or suppliers. It is assumed that each in the chain will be able to recover for the economic loss suffered by his co-contractant so that this loss will ultimately fall on the genuinely defaulting party. So in the *McManus Childs* case it was said that 'If the employer can recover damages the contractor will generally not have to bear the loss since he will have bought from a seller who will be liable . . . and if that seller had in turn bought from someone else there will again be a liability so that there will be a chain of liability from the employer who suffers the damage back to the author of the defect'.

Unfortunately life in the real world is not so simple. It is often the case that the chain has a weak link – the overseas firm with no assets within the court's jurisdiction or the contractor/sub-contractor with no funds. As a result of the recent reversal of the trend of allowing actions in negligence for the recovery of economic loss where proximity and reliance could be established, a break in the chain will normally mean that the party suffering the loss will have no opportunity of recovering it from the true defaulter, unless he has protected himself by an appropriately drafted collateral warranty.

This is even more the case when the parties have constructed their contractual relationships in such a way as to show their exclusive reliance on contractual remedies. In *Greater Nottingham Co-operative Society v Cementation Piling and Foundations Ltd* [1989] QB 71, it was decided by the Court of Appeal that where the employer had taken a collateral warranty from a sub-contractor which was limited to design and selection of materials, but did not extend to workmanship, the employer could not recover financial losses which were due to the way in which the sub-contractors had negligently executed the works. The direct contract in the form of the collateral warranty was considered as being inconsistent with any assumption of responsibility by the sub-contractor, certainly for economic loss, beyond that which he had *expressly* undertaken.

In *Simaan General Contracting Co. v Pilkington Glass Ltd* [1988] 1 QB 758, specialist glass window units had been supplied by Pilkingtons to the installation contractor Feal who were sub-contractors to the main contractors Simaan Contracting for a new building in Abu Dhabi. The colouring of the units was defective and ultimately they were rejected. Simaan then brought an action in negligence against Pilkingtons instead of suing the sub-contractors Feal for breach of contract. The Court of Appeal, in rejecting the claim, took the view that the parties having deliberately formed a chain of contracts, main contractor with the installation contractor and installation contractor with supplier, must be assumed to have contemplated that any claims would be made down the contractual chain and not short-circuited by an action in tort. There was no evidence that Pilkingtons had ever assumed any direct responsibility towards Simaan.

It follows from these cases that in establishing his contractual arrangements the employer, if he wishes to have any rights to recover for economic loss against a party with whom he would not normally have any contractual relationship, e.g. a sub-contractor, must do so expressly in contract by way of a collateral warranty and that he must ensure that the terms of the collateral warranty cover all the obligations of the party concerned.

### Professional advisers

English law has long drawn a distinction between the obligations in contract of a contractor or supplier and the obligations of a professional man. In general the obligations of a contractor or supplier are strict; that is to say they are not based on fault and it is no defence that all reasonable care was taken. If in a construction contract the contractor is responsible for design then, unless the contract provides otherwise, the contractor is strictly liable for design and the works must be fit for the purpose for which they were intended. The obligations of the professional man however, in the absence of any express term in the contract to the contrary, or a warranty which the courts are prepared to imply as a matter of fact, are only to 'carry out the service with reasonable skill and care' (s. 13 of the Supply of Goods and Services Act 1982) or as it has been described in the courts to bring to the task 'the standard of the ordinary skilled man exercising and professing to have that special skill'. The question whether reasonable skill has been exercised or not is a question of fact which in practice largely rests upon whether or not other people in the same profession being persons of skill and experience would have behaved in the same way or not having regard to the state of knowledge existing at the time. This is not however in any way a rule and if, exceptionally, what is common practice in the profession is judged to be negligent then the professional will as it has been put 'pay for the sins of his profession'.

The normal obligation then of a professional man does not extend to guaranteeing a result. If there is to be such a guarantee then there has to be an express term in the contract to that effect, or the court must find on the evidence that the contract includes a term implied as a matter of fact that the professional man is responsible that the works are fit for the purpose intended. Such a term will not be implied as a matter of law where the contracting party is a professional man providing only advice or designs, i.e. without supplying any product (*George Hawkins v Chrysler and Burne* [1986] 38 BLR 36). Nor somewhat more strangely does it appear that even if the professional person in question actually possesses a higher than normal degree of skill is he to be judged by that higher standard. There is apparently no stricter liability than that of 'ordinary' negligence (see *Wimpey Construction UK Ltd v Poole, The Times* 3 May 1984).

However where the design is linked to construction, as in a packaged deal contract, the obligations as to design and construction will be considered as an integral whole and since the object of such a contract is normally to provide the employer with an entire installation capable of achieving a specified result, the liability for design will be based on fitness for purpose regardless of negligence or fault and if such a term is not expressly included within the contract it will be implied (see *Viking Grain Storage Ltd v T.H. White Installations* 3 CON LR 52, following the decision of the Court of Appeal in *IBA v EMI Electronics Ltd & BICC* [1978] 11 BLR 29). While the House of Lords did not expressly decide the point when that case came before them, since reversing the Court of Appeal they found the design to have been negligent, their speeches indicate general agreement with the Court of Appeal on that issue. As regards the position of a consulting engineer employed by the main contractor in such a case to perform the design if he is provided with all necessary information as to the purpose for which the installation is required, then in the absence of any express provision to the contrary a term may be implied in fact in the contract between the package deal contractor and the consultant, that the consultant's design will similarly be fit for the purpose intended without proof of negligence (see *Greaves v Baynham Meikle* [1975] 3 All ER 99). It is to be noted that in the Greaves case the term was implied *in fact* based on the evidence of the intention of the parties.

# The measure of damages

The measure of damages which the employer may be able to recover from the defaulting party will differ according to whether the claim is against the main contractor in contract or against the sub-contractor in negligence to the extent that the employer is still entitled to make any such a claim having regard to the decisions in *Murphy* and *D & F Estates* referred to earlier.

### Damages in contract

The basic principles may be stated as follows:

1   Damages are compensatory and the objective is to put the injured party, so far as money can, in the same position as if the contract had been performed. It follows from this that damages can be recovered for the loss of expectations arising out of or created by the contract. It is on this basis that an employer can – in principle and provided they are not too remote – recover damages for his loss of profits arising from works which do not perform according to specification and not merely for the costs of putting them right.

2   There are two alternative bases of assessment which may be applied in contracts for engineering works. The one has been referred to as the 'difference in value' and the other 'the cost of cure'. In general it would appear that in the event of the contractor failing to perform the work correctly in accordance with the specification the assessment will be on the basis of 'the cost of cure' and this may still be so even if it results in the employer eventually being placed in a better position than he would have been had the original contract

been properly performed. So when a factory was burnt down because of the breach of contract by the contractor the employer recovered the full costs of re-building even although that gave him a new factory. Only where the costs of 'curing' would be exceptionally high, as when sub-standard components have been built into a structure which would require extensive work on it to put the defects right, will the alternative basis of 'loss of value' be applied. In practice if the effect on the installation of the inclusion of the sub-standard components is substantial the two bases may in fact produce largely the same result.

3   The damages must not be 'too remote'. Since the decision of the House of Lords in *The Heron II* [1969] AC 350 (under the name *Koufos v C. Czarnikow Ltd*), the loss must be a 'serious possibility', and it is on that basis that in contract the words 'reasonably foreseeable' must be interpreted. What is a 'serious possibility' will depend upon:

- what the defendant must be presumed as a reasonable man to have known at the time of entering into the contract. In making that assessment it is appropriate to take into account the capacity in which the defendant contracted. So an experienced contractor erecting a block of flats for a property developer must be presumed to know that the employer intended to let them at a profit. Accordingly, if he is late in completion he would be liable to compensate the developer for such loss of profits as were reasonably foreseeable.

- any actual knowledge which the defendant possessed at the time of entering into the contract and on the basis of which he must be presumed to have contracted. This is obviously reasonable in that such knowledge would have allowed him the opportunity of protecting himself against the risk by, say, taking special measures to ensure completion on time, or covering himself by insurance against the consequences of defective design. So if the contractor in the example above was specifically advised by the developer at the time of tendering that the building was for occupation by, for example, foreign embassy staff who would be paying exceptionally high rents, then he would be liable to pay damages based on those rents were he to fail to have the flats ready for occupation by the contractual date.

4   Contributory negligence is not a defence to a claim for damages founded on a breach of a strict contractual obligation. So where a contractor had amongst other obligations undertaken that 'their materials and workmanship would be the best of their respective kinds' the damages suffered by the employer could not be reduced because of any alleged failure by the employer to disregard his own interests (*Barclays Bank plc v Fairclough Building Ltd, The Times* 11 May 1994).

## Damages in negligence

The general rules may be stated briefly as follows:

1   Once negligence has been established then the person responsible will be liable for the damages which are of a type which were reasonably foreseeable or a probable consequence of his act. It is not necessary that the actual detailed circumstances should have been reasonably foreseeable provided that the general category was so.
2   Provided the damages were of a *type* which was reasonably foreseeable then it is irrelevant that the actual *extent* of the damage or loss which occurred was reasonably foreseeable. Compensation is payable in respect of the harm which was actually suffered.
3   In principle the person who has suffered as a result of the negligent action is entitled to be put into the same position – so far as an award of damages can – as he would have been had the negligent act not occurred.

## Distinction between contract and negligence

The main points of distinction are:

1   The 'foreseeability' test in contract is stricter than it is in negligence. In contract it is not a question of 'reasonable foreseeability' as it is in negligence but as 'not unlikely' or 'serious possibility' in the contemplation of the parties. It is the subjective element of the contemplation of the parties in contract which makes the difference. The stricter test in contract is justified because it is always open to the one party to bring to the attention of the other at the time when the contract is made the special risk against which he wishes to be protected. No such opportunity occurs in negligence.

2   The 'foreseeability' in negligence applies to the type, kind, degree or category of loss and not to the actual damage suffered. In contract it is the losses themselves which must have been a 'serious possibility' in the contemplation of the parties if they are to be recovered. It is on this basis that some loss of profits was allowed in *Victoria Laundries v Newman* and vendors of land have been required to pay for the purchaser's anticipated loss of profits for re-development *where they knew that re-development was intended.*

3   Contributory negligence can operate as at least a partial defence where the cause of action is founded in negligence or could be.

4   In contract the award of damages is intended to put the party in the same position, so far as it can, and within the rules as to remoteness, as if the contract had been performed. The award of damages in negligence is intended to put the injured party in the same position as if the negligent act had not happened. So in an action for negligent misrepresentation the damages would be based on what the position would have been had the misrepresentation not been made. In an action in contract for misrepresentation the damages would be based on what the position would have been had the misrepresentation been true.

## Nature and extent of the employer's liabilities

The obligations of the employer in contract will in general be set out expressly in the terms of the particular contracts into which he enters. However there are certain obligations which as a matter of law will be implied and are of particular significance to the state of contract planning. These are:

1   It is an implied term of any construction contract that if the performance of the contract requires the co-operation or action of the employer then the necessary degree of co-operation or action will be forthcoming. It seems doubtful if this particular implication can be negated by the express terms of the contract.

2   Following on from (1) if the employer undertakes to supply drawings, instructions or approvals then there is an implied obligation that such will be given in a reasonable time and so as to enable the contractor to comply with his contractual obligations.

3   Again following on from (1) if the employer undertakes to supply components to a contractor for incorporation into the works there is an implied obligation they will be supplied in time to avoid disruption and delay (*Thomas Bates v Thurrock Borough Council* Court of Appeal 22 October 1975). It was admitted in this by the Council that there was additionally an implied term that the components would be of good quality and fit for their intended purpose.

4   If information is supplied by the employer to the contractor on the basis of which the contractor is to undertake his contract, e.g. information as to ground conditions on which the contractor is to prepare his designs, then it is an implied term that such information should be correct (*Bacal Construction (Midlands) Ltd v Northampton Development Corporation* [1975] 8 BLR 88).

   The attempt is very often made by employers, and indeed by main contractors in dealing with sub-contractors, to limit the scope of application of any such implied obligation by providing that any information given is not guaranteed and it is the responsibility of the recipient to check it for himself. The following comments are made as to the legal effects of such attempts:

- If the facts of the case show that the tenderers were intended to rely on the information provided as regards the soil conditions, and did so rely and thereby suffered loss because the information had been negligently prepared, then the consulting engineers who prepared such data may be liable to the contractor under the principle established by the *Hedley Byrne* case. This may be so even if there is a disclaimer in the bidding documents which protects the employer but not specifically the consultants, i.e. any disclaimer clause will be construed strictly against the party imposing it. In deciding upon whether or not it would be reasonable to impose such a duty in tort, the court may take into account the practicalities of the tenderer's ability to undertake any investigations for himself. In the Canadian case of *Edgeworth Construction v ND Lea & Associates and Others* [1993] 66 BLR, the Canadian Supreme Court took into account, in finding that a duty of care existed on the part of the consultants, the fact that the bidders had about two weeks in which to file their tenders and the consultants had spent two years on the preparation of the engineering design and information.

  Although there was no disclaimer protecting the consultants in that case it is thought that, if on the same facts an English court reached the same conclusion on the existence of a duty of care, then it seems unlikely that they would uphold the validity of any such disclaimer under the Unfair Contract Terms Act.

- If the misrepresentation is made fraudulently which means either (a) knowing it to be false, or (b) without belief in its truth, or (c) recklessly not caring whether it is true or false, then no disclaimer clause will act to protect the person making the misrepresentation and this applies whether the misrepresentation was made by the employer or his agent. For this purpose the House of Lords have said principal and agent are one (*S. Pearson Ltd v Dublin Corporation* [1907] AC 351).

- If the information given amounts to a misrepresentation

then under the Misrepresentation Act 1967, as amended
by the Unfair Contract Terms Act 1977, the employer will
be liable to the contractor in damages unless he can show
that 'he had reasonable grounds to believe and did
believe up to the time that the contract was made that the
facts represented were true', and further that the
disclaimer clause in the contract satisfies the
requirements of reasonableness as stated under s.11(1) of
the Unfair Contract Terms Act. The important point is
that in each instance the burden of proof is on the
employer both as regards establishing his belief in the
factors and shown that the clause was reasonable.

The employer will be held liable under the Act for a misrepresentation made by his agent which would cover the case where
it was made by his consultants. It would not therefore appear to
be a defence for the employer to argue that he had employed
and relied on professional advice. Nor would it be a defence for
him to argue that it would have cost too much time and money
to establish the truth (*Court of Appeal in Howard Marine and
Dredging Co v A. Ogden & Sons (Excavations) Ltd* [1977] 9 BLR 34):
'In the course of negotiations leading to a contract the statute
imposes an absolute obligation not to state facts which the
representor cannot prove he had reasonable grounds to believe.'
As regards establishing that the disclaimer clause is
reasonable then it is considered that the Court would take into
account the complexity, time and cost of investigating and
verifying the data provided, within the period allowed for
tendering together with the significance of the data to the
tenderer and would be likely to hold that, unless the
investigations needed were of the simplest, a clause seeking to
establish a total disclaimer did not satisfy the test. Thus on a case
decided before the Act it was held to be unreasonable to require
a tenderer who had seen two trial holes to search an overgrown
site to find three others of which they were not aware! (*Bryant &
Son Ltd v Birmingham Hospital Saturday Fund* [1938] 1 A11 ER 503
at p. 21).
It is recognized that a contractor who has under-priced a job

for reasons unconnected with the data supplied may nevertheless seek to use any inaccuracy in such data as a means upon which to found a claim. However the fact that such a possibility exists provides in the author's view no justification for seeking to impose upon tenderers obligations with which in practice they clearly cannot comply. Further it must be to the employer's advantage that the contract starts off with the contract price based as securely as possible upon the conditions which will actually be met when the work is performed. Whilst it is in the nature of soils investigation work that there can be no guarantee that this will be the case there is surely everything to be said for such work being carried out with the maximum of care and to an extent sufficient to reduce so far as practicable the possibility of unpleasant and expensive surprises.

## Employer's liability for damages

The general issues relating to damages discussed earlier in respect of the contractor apply with equal effect to the employer but there are certain specific issues which may arise out of a breach of contract by the employer which need noting. These are:

1 Where the contract uses the term 'direct loss/expense' or 'direct loss/damage' as in JCT 80 clause 28.2, then this will be interpreted as equivalent to the damages which would follow directly from a breach of contract and would include therefore the contractor's loss of profit (*Wraight Ltd v P. H. & T. (Holdings Ltd* 13 BLR 26). The court in fact allowed 10 per cent for establishment charges and profit which the contractor would have earned on the contract had it not been determined and 12 $^1/_2$ per cent for a proportion of his overhead costs attributable to the contract.

2 Head office overheads in terms of additional managerial expenses required in attending to the problems caused by the employer's default can be recovered in a claim for damages provided they are properly quantified. It is not sufficient merely to add a percentage to the direct costs (*Tate & Lyle Food and Distribution Ltd v GLC* [1982] 1 WLR 149).

3   If under the terms of the contract, e.g. JCT 80 clause 26, as a result of the actions of the employer the contractor is prevented from utilizing his resources on other work, and can prove that he could have done so, then the anticipated loss of profit on such other work is recoverable for the period when he was so prevented (*Peak Construction (Liverpool) Ltd v McKinney Foundations Ltd* [1970] 1 BLR 111). Under the 6th edition of the ICE conditions the term 'costs' is defined as 'all expenditure properly incurred or to be incurred whether on or off the Site including overhead finance and other charges properly allocable thereto but does not include any allowance for profit'. However under a change from the 5th edition the contractor under clause 42 (delay by the employer in giving possession of the site) is entitled to his additional costs together with an 'addition for profit'.

4   The term 'direct loss and/or expense' under the JCT form of contract includes interest or financing charges and these will be assessed on the same basis as the bank assessed such charges on the contractor, i.e. with periodical 'rests' at which point the interest outstanding was added to the principal (*F.G. Minter Ltd v Welsh Health Authority Technical Services Organisation* [1980] 13 BLR: *Rees & Kirby v Swansea City Council* [1985] CILL 188).

5   The use of the expression 'consequential loss' in a clause seeking to limit liability will not prevent the recovery of those damages which flow directly and naturally from the breach and these will include loss of profit (*Millar's Machinery Co. Ltd v David Way and Son* [1934] confirmed by the Court of Appeal in *Croudace Construction Limited v Cawoods Concrete Products Ltd* [1978] 8 BLR 20). The term 'consequential' means 'merely consequential' and 'something not the direct and natural result of the breach'. What this leaves the term 'consequential loss' to mean is not altogether clear but the probable answer is those damages for which the defendant would be liable because of some special knowledge which he had at the time of entering into the contract (see p. 61) and on the basis of which it is presumed that he did enter into the contract.

## Liability of the employer in negligence

As between the employer and the main contractor the question of liability in negligence is not one which should often arise. This is simply because as stated by the Privy Council in *Tai Hing Ltd v Liu Chong Hing Bank* 'their Lordships do not believe there is anything to the advantage of the law's development in searching for a liability in tort where the parties are in a contractual relationship. This is particularly so in a commercial relationship' (*Weekly Law Reports* 2 August 1985). In so far as the parties have set out in some detail their respective rights and obligations within the contract between them then it is to the contractual terms that reference should be made when any dispute arises.

It would seem that an employer would not be liable in tort for the acts of his architect or engineer if, as a professional man, he was acting as an independent consultant. He would then be in the same position as any independent contractor. In many instances architects or engineers are not independent consultants but employees of the employer and in these circumstances the employer could be vicariously liable for their negligence. Further, even when the architect or engineer is an independent consultant, the influence increasingly exercised by administrative and financial departments in the employer's organization may result in it happening that:

the exercise (by the architect/engineer) of his professional duties is sufficiently linked to the conduct and attitude of the employer that he becomes the agent of the employers so as to make them liable for his default. In the instant case the employers through the behaviour of the council and the advice and intervention of the town clerk were to all intents controlling the architect's exercise of what should have been his purely professional duty. In my judgement this was the clearest possible instance of responsibility for the breach attaching to the employers.
(*Rees & Kirby v Swansea City Council* in the High Court)

Although at one time it was thought that an architect or engineer owed a duty of care to a contractor who would be affected if the architect/engineer were negligently to undercertify the value of his work, this now seems doubtful following the decision of the Court of Appeal in *Pacific Associates v Baxter* [1990] QB 993. The contract in that case contained an arbitration clause and a disclaimer of the defendant's liability. In essence the decision seems to have turned on the structure of the contractual relationships between the parties and the fact that the contractor could claim against the employer in arbitration.

Assuming the contract contains an arbitration clause, then it would seem that a claim by the contractor against the architect/engineer for under-certification would only be likely to succeed if the architect/engineer were to have acted deliberately in contravention of the contract with the intent to deprive the contractor of money to which he knew that the contractor was entitled. This was the view of the Court of Appeal in *Lubenham Fidelities v South Pembrokeshire DC* (see 6 Con LR at page 114).

## Pre-contractual discussions

These may affect the relationships between the parties because of:

1 Representations.
2 Collateral warranties.
3 The issue of Letters of Intent.

### Representation

The problem of misrepresentation has already been discussed in relation to one of the areas in which it is most likely to arise, namely the giving of data relating to site and soil conditions, and the principles set out there are of general application. However it is worth emphasizing that a representation is any statement of fact made by one party to the other before the contract is made and which induces the person to whom it is made to enter into the contract. The representation must be as to a matter of fact and not just an assertion of opinion. However if the opinion is expressed by someone having or claiming special knowledge or skill in relation to the matter in question, or if by implication it is founded on facts, then it will still be treated as a representation. In practice therefore, when inviting tenders an employer should be extremely careful as to the data which he provides to the tenderers. Unless the matters are ones which it is impractical to expect the tenderers to find out for themselves, it is far better simply to make it clear that it is their business to find out the information they require in order to bid. It is also a point which needs watching when conducting bidders conferences, or answering bidders questions.

The general position as to liability for misrepresentation can be summarized briefly as follows:

1   If the representation is fraudulent (for the meaning of this see p. 65), then the remedy is damages and recission.
2   If the representation is made negligently, i.e. carelessly and in breach of a duty of care, then remedy is damages. The duty of care generally arises out of a professional relationship, e.g. architects and engineers, and will be owed to persons such as contractors who act in reliance on their statements in circumstances when it was reasonable to expect that they would do so. It may however arise out of a commercial relationship if one party acts on the special knowledge and expertise of the other and it should have been foreseen that he would do so. Thus when a sales manager on his employer's behalf provided a tenant of a petrol station with a statement as to potential turnover on which the tenant relied, it was held that the company owed the tenant a duty of care (*Esso Petroleum Co. v Mardon* [1976] QB 801).
3   The statutory liability as already discussed under the Misrepresentation Act 1967.

**Collateral warranties**
A collateral warranty normally arises when an undertaking is given during contractual negotiations as to some matter, which is intended to have contractual effect, but which is not included within the contract terms, and indeed may even be in contradiction to them. In the usual course of negotiations

between the parties statements will be made and requests for information answered and it is a matter of fact to be determined in each case, whether or not looked at objectively there was a clear intention on the part of the parties that such statements or responses should constitute contractual obligations. The attitude of the courts in general is that the existence of a collateral warranty is to be the subject of strict proof (see the comments of Viscount Dilhorne in *IBA v BICC* [1980] 14 BLR 1).

In the *Esso Petroleum v Mardon* case the Court of Appeal also held that there was a breach of a contractual collateral warranty that the estimate of turnover had been prepared with due care.

Following the *Murphy* and *D & F Estates* decisions, collateral warranties have now assumed a far greater importance. Since effectively the employer has no remedy in tort against a defaulting sub-contractor for defective work, the only way in which he can protect himself is to obtain from the sub-contractor a collateral warranty (see Chapter 16, p. 288–91).

### Letters of Intent

The best advice which can be given to any employer contemplating the issue of a Letter of Intent is 'don't, or if there are compelling commercial reasons then exercise the greatest of care'. In so far as the Letter merely expresses an intention to award a contract *and nothing more is either said, written or done*, then since the Letter on its own creates no contractual obligations on either party, and is of no binding effect, it can be argued that the employer had done himself no harm except to weaken his negotiating position when it comes to the contract. However, the purpose of issuing the Letter of Intent is almost always that *something is to be done* for which the contractor wants the assurance of payment and once the Letter has been written further actions and correspondence will follow. Where this is the case and the contractor actually performs preliminary work for the employer then the employer will be liable to pay for it even if the project never actually proceeds (*Turiff Construction Ltd and Turiff Ltd v Regalia Knitting Mills Ltd* [1971] 9 BLR 20).

The alternative situation can, from the employer's viewpoint, be even worse: where at the employer's request work is started and *completed* on the basis only of a Letter of Intent because the parties never finally agreed a contract. The contract work having been performed at his request the employer is bound to pay for it on a quantum meruit basis but no contract ever having been concluded the contractor is under none of the normal obligations for quality of work, delivery on time etc., which would either be implied by law or form part of the express contract terms (*British Steel Corporation v Cleveland Bridge and Engineering Co. Ltd* 21 December 1981).

# Conclusions

From this brief survey of certain legal issues the following conclusions relevant to the subject of contract planning can be drawn:

1   The employer is in the best *legal* position as regards minimizing his own risks and placing the maximum liabilities on the contractor by the placing of a turnkey contract. However he must be sure that the turnkey contractor has the necessary financial resources to support the responsibilities he is accepting; that he is worth 'powder and shot' if it should ever come to legal action or even the threat of it.

2   That if the employer wishes to separate out design from construction or manufacture, then he should seek from the designers guarantees that their designs will produce the results intended, if construction/manufacture is properly executed, and give to the designer the responsibility for supervision of construction/manufacture. He should not be content to rely on the traditional obligation of the consultant to use reasonable skill and care.

3   The methods of contracting referred to in Chapter 2 as 'client co-ordinated' and 'management contracting' impose on the employer the liability towards each contractor of the consequences of the default of any other. In separating out the contracts the employer should seek to minimize the number into which the project is divided and should

consider carefully the extent to which he can obtain indemnities enforceable in practice, at least from firms responsible for key areas of the work.

4 If the employer either directly or through another designer/contractor/supplier has accepted the responsibility for the supply of drawings, data, components or other services/facilities, then he should select either the company supplying the item/service in question or the one receiving it, to be responsible for its suitability, quality and delivery to time. Clearly if the employer is supplying the item/service directly himself then, if at all practicable, he should place that responsibility with the recipient.

The employer must identify and place responsibility for positively managing the interface on the firm most appropriate for the task and be sure again they have the financial backing should things go wrong. However it is to the employer's legal advantage to seek to reduce to a minimum the provision of such items/services. Some will be unavoidable, as for instance soil investigation reports when inviting competitive tenders for construction, but supply of free-issue items rarely is – and should be – avoided.

5 It is to the employer's legal advantage to place the responsibility for sub-contractors firmly in the hands of the main contractor and to play no part in their selection or to know of the terms on which they have been employed. It is recognized that with the complexity of modern contracts, and the extent of sub-contracting which takes place, there may be technical or commercial reasons why the employer does wish to get involved, but he must be very careful not to dilute the responsibilities of the main contractor while at the same time ensuring that he has an effective remedy against an important sub-contractor through the use of a collateral warranty.

6 If for commercial reasons the employer wishes to make use of the nominated sub-contractor/supplier system then he should ensure that he has direct contractual rights against the nominated firm in the event of their failure to perform, and not rely on the possibility of being able to prove either

negligence or breach of an implied collateral warranty. The JCT have to their credit recognized and sought to tackle most of the problems of nomination (except re-nomination see post p. 292) but a study of their recommended procedures and forms shows the complexities into which the supposed advantages of nomination lead. Again the employer should satisfy himself on the financial resources of the firm in question and if necessary insist on the provision of a bond.

7   The employer must be conscious of the liabilities which he is accepting towards his contractors either in contract or negligence and whether due to his own default or that of his consultants. While as suggested above he should seek to lay these off as far as he can on others who possess the requisite financial resources, he should assess the residual risks and liabilities which remain with him, and what provision he needs to make in his financing to cover them.

8   The contract plan should be formulated as a whole in a way which will be clear and definite and avoid the need for extensive pre-contract debates at which it is only too easy for potentially damaging representations to be made, and also the uncontrolled issue of Letters of Intent. If these have to be used on occasions some suggesting wording is given on p. 157.

# Part Two
# TENDERING AND PLACING THE CONTRACT

# 4 Competitive tendering

One of the matters to be dealt with in the contract planning exercise is the method by which the contractors for the project are to be chosen. The methods most commonly used are set out below:

1. 'Open' competitive tendering for the whole contract price by advertisement.
2. Competitive tendering for the whole contract price from a selected list – selective tendering.
3. Competitive tendering from a selected list but limited to rates and fees.
4. Selection of single contractor with whom to negotiate.

## Open competitive tendering

This is one of the methods for public works contracts which is prescribed by the European Directive on Public Works Contracts which was issued in its consolidated form as Council Directive 93/37 of 14 June 1993 and in the Public Supplies Directive 93/36 of 14 June 1993. It is also referred to in the Utilities Directive of 14 June 1993. Its use in the UK appears to be very limited. According to figures published by Euro-Bid Watch in 1993 only 2 per cent of tenders for public works contracts in the UK were open.

While it is clear that the Works Directive applies generally to building and civil engineering work there is a difficulty with contracts on a turnkey basis for the design, supply and installation of plant and equipment. Such contracts in normal English practice would be treated as works contracts but not so apparently under the Directives. It would appear that with mixed contracts for supply and installation whether the contract is to be treated as a works contract or a supply contract depends on the respective values of the supply portion and the erection work. If, as would usually be the case, the value of the supply exceeds that of the installation, the contract would be subject to the Supplies Directive and not the Works Directive (see Trepte paras. 401, 403 and 405). Nevertheless it is to be noted that an urban waste disposal plant was apparently considered as subject to the Works Directive – see Case 199/85 quoted by Trepte on p. 126.

The importance of the distinction is twofold. First there are different threshold values for the application of the respective Directives. Secondly the provisions of the Supplies Directive relating to technical competence – see further below – are totally inadequate to deal with engineering contracts where the contractor is responsible for the design, supply, installation and commissioning of a complete plant.

To some extent the EC Works and Supplies Directives have sought to overcome one of the main objections to the use of the open method, which is that the purchaser having received bids from a wide range of contractors of differing skills, abilities and financial resources is placed in an extremely difficult position when it comes to awarding the contract. The purchaser is allowed to exclude firms if they fail to meet the criteria established in the Directives relating to general suitability, financial and economic standing and technical competence (Arts 24–27 of the Works Directive and Arts 22 and 23 of the Supplies Directive).

For details as to these rights of exclusion see in particular *Public Procurement in the EEC* by Trepte, CCH Europe 1993. It is sufficient to note here that the Directives provide lists of the evidence by means of which the purchaser is to establish whether or not the contractor does meet the criteria. It seems that while these lists are not exhaustive in relation to economic and financial standing, they are so in regard to general suitability and technical competence. So in a case to which Trepte refers the Belgian authority rejected the lowest tender in favour of the next lowest on the grounds that the workload of the lowest bidder was in excess of the level laid down by Belgian rules. The court ruled that this value of work rule was a useful measure in determining the contractor's ability to undertake further work and was not contrary to any EEC rules.

For the important issue of technical competence it appears that the five references set out in Art. 27 of the Works Directive and Art. 23 of the Supplies Directive are exhaustive and are intended not just as references but to establish the only criteria upon which technical competence can be judged. While, as Trepte points out, a purchaser can set the level of competence required, he cannot require standards other than those contemplated by the references in the Directives. While the criteria are reasonable for the Works Directive they are not suitable for design, supply, install and commission contracts and this could create problems for the purchaser.

In the only case to come before the UK courts so far a housing authority was held to be entitled to take into account criteria relating to compliance with health and safety matters, on the grounds that technical capacity to carry out works competently includes the ability to carry them out with due regard to the health and safety of those the contractor employs and the general public. The decision seems eminently sensible. In fact the authority did have evidence available to it as to the contractor's safety record on other contracts (*General Building and Maintenance v Greenwich Borough Council*, The Times 3 March 1993).

Even with the provisions in the Directive allowing for a certain exercise of judgement by the purchaser as to the contractor's competence, this still does not remove the other objections to open tendering. First, knowing that there will be a large number of firms submitting tenders, some of whom will be willing to take chances and submit 'cut price offers', the more competent contractors are likely to be deterred from putting the necessary time and effort (which both cost money) into the preparation of their own tenders and may limit themselves, not unreasonably, to 'cover' prices. They may even decide not to tender at all. Second, the process of screening all the offers received, taking up references and checking the tenderer's financial resources and technical ability is an extremely costly and time-consuming process. Unless it is carried out thoroughly and competently the purchase will end up accepting a low price tender from a firm which is not suited to carrying it out, and while the initial price may be low, the final cost (including the cost of delays, claims and making good) is likely to be substantially higher.

## Selected list

Here the purchaser initially selects a short list of firms whom technically and commercially he considers suitable to undertake the work in question. Normally it can be expected that the purchaser either from the experience of his own commercial and engineering departments or with advice from his consultants will be able to select the firms on his own initiative. On international tendering however it is common for a public invitation to be issued for firms to pre-qualify. This means that firms interested in bidding for the project can inspect the bidding documents and submit details of their competence and experience to undertake the work involved. They will be required to complete a questionnaire detailing similar work previously carried out, numbers and qualifications of their professional staff, a statement of their financial assets with a copy of their latest balance sheet, particulars of their manufacturing facilities etc. A useful guide to the preparation of pre-qualification forms is the Standard Pre-Qualification Form for Contractors issued by the International Federation of Consulting Engineers (FIDIC), PO Box 17334, 2502 CH, The Hague, Netherlands. For manufacturing work this will need to be supplemented by requests for information relating to relevant manufacturing capacity and proportion already booked and quality assurance and quality control procedures.

As the replies are received they should be recorded in a register.

Where the contract falls within the scope of the Public Works Directive, the Supplies Directive or the Utilities Directive of 14 June 1993, then this method which is referred to there as the 'restricted procedure' may be used as an alternative to the open procedure and there is no restraint on the purchaser as to which he chooses. It is not proposed to go through the procedures – for details see Trepte or other standard texts. They necessarily involve the issue in the *Official Gazette* of the EC of a notice asking for requests to participate and giving particulars of the works and the intended contract.

For public works and supplies contracts the purchaser may, when using the restricted procedure, exclude any firm from the list of those invited to tender by reference to the criteria referred to above under the open procedure. Indeed the case of *GBM v Greenwich Council* referred to above was under the restricted procedure. The actual selection of those to be invited to tender is then to be made on the basis of their past performance and the other information obtained relating to the criteria for qualification without any discrimination between firms in different member states.

The Utilities Directive is more relaxed. The utility can select according to 'objective criteria and rules which they lay down and which they make available to interested contractors'. Again of course there must be no discrimination on grounds of nationality.

In addition to the restricted procedure there is also under all the Directives the negotiated procedure under which 'the purchaser consults contractors of his choice and negotiates the terms of the contract with one or more of them'. The purchaser must state in his notice in the Official Journal which he intends to use. In practice for public works or supplies contracts the purchaser's ability to use the negotiated procedure is extremely limited. For details see Art. 7 of the Works Directive and Art. 6 of the Supplies Directive.

The Utilities Directive, however, gives the purchaser an unrestricted choice as to which procedure to use provided only that there is a call for competition. Exceptionally the negotiated procedure may be used without a prior call for competition – for details see Art. 20(2).

The advantage which the utility certainly appears to obtain by the use of the negotiated procedure in competition is that they can then enter into post-tender negotiations and eliminate progressively those firms whose bids are not acceptable. Under the restricted procedure this would not appear to be possible since the Commission have stated that with either the open or restricted procedures negotiations with tenderers are ruled out on fundamental matters relating to their tender which would distort competition such as price.

One very important point to note in relation to the operation of the procedures is that, assuming the purchaser wishes to make his award on the basis of the 'most economically advantageous offer', as opposed to the lowest price, he must set out in the notice appearing in the Official Journal of the EC, details of the objective criteria which he intends to take into account when making his award. These criteria must not be such as to discriminate against any tenderer from a third country, e.g. one which referred to an obligatory requirement to use a percentage of local labour. The Directives give as some examples:

price
delivery or completion date
running costs
cost effectiveness
profitability
aesthetic and functional characteristics
technical merit
quality

after-sales service
spares.

They are indeed the type of criteria which a competent purchasing organization would use whether in the public sector or not.

From an analysis of the particulars thus submitted the purchaser and his advisers are able to select the short list from whom tenders will be invited.

In the exercise of selecting the short list four points in particular need stressing:

1   The selection needs to be done positively, not through the time-honoured principle of 'Buggins' turn'. On a large job the prospective bidders should be interviewed to assess their interest and suitability for the particular job at the time in question.
2   Like must be matched against like. It is no use putting the local builder in competition with a major national contractor, nor asking Harrods to tender against Woolworths. The list should be related both to the size of the job and to the quality which the purchaser wants and, equally important, is prepared to pay for. It is considered that this would not offend either the restricted or the negotiated procedures, provided it was clear that the actual selection was made objectively and without discrimination.
3   The operative word in describing the list is 'short'. Long tender lists are a menace. The tenderers get to know the list is long and some, perhaps the best, will lose interest. The purchaser's task in tender appraisal is made more arduous. Worst of all is the waste of time and money in the contractors' tendering offices, or the pernicious practice, which long tender lists serve only to encourage, of 'cover' prices. In the restricted procedure under both the Works and Supplies Directives it refers to the number 'being determined in the light of the nature of the work to be carried out. The number must be at least 5 firms and up to 20. In any event there must be enough to ensure genuine competition' (Art. 22(2) of the

Works Directive and Art. 19(2)). The author would entirely agree with the last sentiment but not with the idea that there could be as many as 20 bidders. It would be more realistic to think of 5 to 8. For utilities there is no such restriction. They need only base the number 'on the objective need to reduce the number of firms to a level which is justified by the need to balance the particular characteristics of the contract award procedure and the resources required to complete it' (Art. 31(3)). This is a welcome confirmation that tendering costs time and money both to the firms involved and to the purchaser in his task of evaluation. Again there is reference to the need for ensuring competition.

4   The selection should be done objectively by a two-stage process the details of which should be established *in advance* of the issue of the call of pre-qualification. The first stage of the process is that in which firms are eliminated from further consideration because they fail to meet certain minimum criteria. Typically such criteria could be:

- Lack of recent technical experience of similar-class work. It is for that reason vital to obtain particulars of work of the type in question executed within the last, say, five years.
- Inadequate financial resources to support the project. This could be judged by reference to turnover, profitability, level of issued capital and willingness of banks to supply necessary level of credit and bond support.
- Lack of management resources which could be made available from within the company.
- On projects overseas, lack of suitable joint venture partner or inexperience of working in the country concerned.
- On a project involving design and manufacture, lack of design and/or manufacturing facilities of the type required and/or of complying with the necessary quality standards.

As referred to above the establishment of objective criteria in advance and their inclusion in the notice in the journal are essential when tendering under the public procurement or utilities rules.

Once the list has been selected the procedure within the purchaser's office concerned commercially with inviting tenders should be as follows:

1  It should be established that all the firms selected are interested and willing to tender.
2  A realistic period should be assessed for tendering; within reason, the longer the better. All of the Directives establish minimum time limits for the period for tendering which for works contracts are to be extended if visits to site are required or there is voluminous documentation to be studied. In general terms under the restricted or negotiated procedures for public works the minimum is 40 days (26 if a prior information notice has been published) and for utilities 3 weeks, unless there is an agreement between the utility and the bidders otherwise in which case the period must be the same for everyone. In practice since site visits and voluminous documentation are the rule and not the exception these limits ought to be increased.
3  A check should be made to ensure that by the date for issue of tenders all the information required will in fact be available. It is no use, for example, finding out at the last minute that a soil survey is needed. If one is required it should be put in hand straight away.
4  The appropriate general conditions of contract should be selected and consideration given to the following points:

    •  whether any modifications are required – for example is the purchaser willing to accept the extensive limitations on the contractor's liabilities under the MF/1 Conditions (see further p. 323)?
    •  if any special conditions are required – for example, the contractor to comply with works safety rules; prohibition against 'poaching' of the purchaser's own labour; the

long-term availability of spares for key items of equipment.

- any blanks in the conditions which it is necessary to complete – for example, percentages of contract price payable on interim certificates and to be held as retention money; amount of liquidated damages for delay and maximum defects liability period if none stated in the general conditions. Many conditions of contract, e.g. ICE 6th edition, have an appendix Part I of which is to be completed prior to invitation of tenders.
- where it is intended to nominate any sub-contractors and suppliers, whether the employer wishes to obtain from such sub-contractors and suppliers a direct warranty – see p. 291.

5  The specification should be examined to see that it describes accurately and comprehensively the work which the contractor is to be required to perform; also that it does not contain anything which is contradictory to the remainder of the invitation to tender including the general and special conditions of contract. Repetition between the documents should also be avoided.

6  The form of tender has to be prepared. This may be quite simple on a standard building and civil contract as the prices will be shown in the bills of quantities. On contracts for the design and supply of plant and equipment or process plants, however, a more detailed form is required which might contain sections as follows:

Section    I    Tender declaration.
         II    Schedule of prices.
       III    Programme.
       IV    Condition of contract.
        V    List of principal sub-contractors and suppliers.
       VI    Management chart showing head office and site supervision proposed.

7  Instructions to tenderers must be prepared, which should

contain clear and detailed instructions as to what work the contractor will have to carry out and how the tender is to be completed. The tenderer's attention should be drawn to any unusual and vital points, and the rules on which the invitation is issued must be made clear. Where the employer is inviting tenders on his own conditions of contract or has modified one of the well-known standard forms, and either his own conditions or the modifications contain some clause which is unusually onerous on the contractor, then it is important that the tenderers' attention is drawn specifically to that clause. Failure by the employer to do so could result in the clause not being considered by the courts as forming part of the contract according to the judgement of the Court of Appeal in *Interfoto Picture Library v Stiletto Visual Programmes Ltd* reported in *The Times* on 14 November 1987. The better the instructions to tenderers and the clearer the form of tender are, the less time contractors will have to spend on their interpretation, the more time they will be able to give to their bid, and the better the offer which the purchaser will receive.

One or two points on the form of tender need amplification as follows:

*Schedule of prices*   The extent to which itemized prices are called for needs watching. It is easy to be over-enthusiastic on this point, but it is suggested that a bill-of-quantity approach to plant contracting is quite out of place and may be positively misleading. It also involves the tenderers in a great deal of unnecessary expense.

*Programme*   The purchaser should state clearly what he wants, and this should definitely not be 'as soon as possible'. The tenderer should be asked to give his own more detailed programme and be instructed to indicate on this any periods which the purchaser considers critical – for example, availability of foundation loads. On contracts of any substance it is suggested that the purchaser should call for a preliminary critical path network to be submitted as part of the tender.

*Conditions of contract*   This should contain the amendments to general conditions, special conditions and variables, damages for delay, for example.

*List of principal sub-contractors and suppliers*   See Chapter 16 on sub-contracting.

*Management chart*   Successful execution of the contract depends upon the degree of concerted effort put into it by the contractor. This in turn depends directly on the extent of management resources allocated. If the contract manager or engineer is trying to do this contract and many others, the proper concentration of effort cannot be forthcoming. The purchaser wants to know therefore, in appraising the tenderer's offer, what he is getting not only in design and materials, but also in management resources, and how much in full-time and how much in part-time. On a major contract the purchaser should ask for the contractor's senior staff, who will be full-time, to be named and their experience and qualifications listed.

## Fees and rates

It is clearly to the purchaser's advantage to tie down the purchase price in as firm a manner as possible before contract award since in any negotiations on the price post-contract the negotiating advantage will tend to lie with the contractor. In most situations it can be expected that the purchaser will be able to give the tenderers enough information to enable them to price the whole tender on a realistic basis even if this means that the price for the building and civil engineering work has to be either wholly or in part on approximate quantities and re-measurement (see p. 225) or plant installation on the basis of man–week rates (see p. 231).

However, situations will arise when because of the urgency of the project, the need to start work and the lack of design data, it is to the purchaser's overall advantage to appoint a contractor, or for a contractor to appoint a sub-contractor on less than a total firm price. One of the most frequent of such situations is when the critical path analysis shows the need for collaboration between the purchaser's and contractor's designers (or the contractor's and sub-contractor's), and for certain long lead items which can be identified early to be placed on order before the completion of design work to a stage at which a totally firm price can be given.

In these circumstances the purchaser or contractor should seek to obtain tenders on the following lines:

1  A price for the tenderer's profit and commercial overheads, preferably as a lump sum but failing this as a scale of percentages on the final cost which reduces as to the cost increases.
2  Rates inclusive of all costs up to and including departmental overheads for the tenderer's design and management staff.
3  A fee for material procurement, again preferably a lump sum but if this is impractical then a percentage on the nett cost to the tenderer for bought-in items.
4  If the tenderer will be involved in installation then rates for his staff and labour inclusive of site overheads.

If the tenderer is going to be involved later in manufacture or fabrication in his own works then he could be asked to give typical budgetary costs for work of a nature similar to that which is envisaged. The quoting of overhead and labour rates for manufacturing work is largely meaningless. High shop-overheads may reflect inefficiency and high total manufacturing cost but they are just as likely to mean that the firm's manufacturing processes are largely automated with consequently a very small labour cost on which the overheads can be recovered.

It is recognized that making a comparison of rates charged by different firms is difficult because the base normally varies from firm to firm and because such comparisons cannot take into account differences in expertise and productivity. The first difficulty can be overcome by specifying in detail the items to be included in the rates and the basis of charging (for details see p. 230). The second difficulty cannot be so easily overcome. Budget estimates prepared by tenderers have a reputation for being notoriously unreliable. This is not always the tenderer's fault since the primary reason for using the form of pricing under discussion is that the employer has not finally made up his mind what he wants and more often than not when he does so he finishes up by wanting more either in terms of quantity or quality than that with which he started. The interesting point does arise as to what duty of care, if any, the tenderer owes to the employer in preparing his budget, or alternatively, can such a budget constitute a misrepresentation under the Misrepresentation Act 1967?

It is well established (see *Nye Saunders & Partners v Bristow* [1987] in the Court of Appeal referred to in *Building Law Monthly* for August 1987 and the Canadian case of *Fidias Consulting Engineers v Thymaras* in the same journal for May 1986) that an architect or engineer certainly owes a duty of care to his client and can be sued in negligence if his estimates of cost have been negligently prepared. Equally, remembering the case of *Esso Petroleum v Mardon*, referred to on page 71, a special relationship giving rise to a duty of care can exist between buyer and seller where the buyer is seeking information of which the seller has special knowledge and which the seller knows will be used by the buyer in making his decision. In principle, therefore, it seems that a tenderer who knew the employer was relying on the information being provided, which he alone was in a position to give, could be liable if the estimate was prepared negligently.

Also if the estimate was based on facts which were plainly not true, and this was known to the tenderer's staff at the time, then, since the purpose of preparing the estimate was to obtain business, the case appears to fall within the decision of the Court of Appeal in *Howard Marine and Dredging Co. v A. Ogden* referred to on page 66, and the employer could have a claim in damages under the Misrepresentation Act 1967.

It is suggested therefore that firms preparing budget estimates as part of their tenders should not take their task too lightly and in particular should not base them upon assumptions which are clearly unrealistic. As for the employer, he should always insist on having the right to question those who prepared the estimate and not allow himself to be forced to deal at long range through the contractor's sales or commercial staff. He should enquire as to the basis on which the figures have been prepared and how they relate to the programme. In short, he should force the tenderer to provide the justification for the budget so maintaining the legal and commercial position which he could lose were he to show that he was substituting his own judgement for that of the tenderer.

Utilizing this method the form of tender can be simpler but must include:

1   The conditions of contract, so that the tenderer can assess any special risks; for example a substantial liquidated damages clause on delay.
2   Programme envisaged, so that the tenderer can assess roughly the period over which he will be committing his resources.
3   The anticipated total scope of the contract, so that the tenderer can assess the total level of the resources he will need to allocate and so the level of return he will expect to receive.
4   Requirement for the tenderer to provide a management chart.
5   Clarification of whether the purchaser's intention is to continue on the basis of fees and reimbursement of costs according to the agreed rates and how in that event other costs, e.g. for bought-in materials, are to be agreed. Alternatively if the purchaser intends that at some stage the tenderer will be required to negotiate a lump sum for the work then the method of negotiation should be stated together with the detail of his pricing which the tenderer will be required to disclose and the extent to which, if at all, any of the rates or percentages originally tendered could then be reopened by either party. This is most important if later serious misunderstandings between the parties are to be avoided.

# 5 Single tender negotiation

For various reasons it may be necessary at times to negotiate with a single contractor. This may be due to the need for speed, because the firm is sole licensee for the equipment or process desired etc. It is suggested that the preferred method of handling such negotiations, for other than items of proprietary equipment, would be as follows:

1   Advise the firm with whom it is proposed to negotiate that it is the intention to proceed on a single-tender basis with them if they are willing to co-operate. Under no circumstances should an attempt be made, by sending out a formal invitation to tender, to deceive the contractor into believing he is tendering in competition. To do this would destroy at once any confidence or good faith as between purchaser and contractor and seriously prejudice the purchaser's chances of future successful negotiations.
2   Agree with the contractor on the basis of negotiation.
3   Confirm the agreed basis of negotiation in a letter to the firm with instructions to them to prepare their specifications and their firm prices. State in the letter that if the negotiations fail

or the work is not proceeded with, then the contractor will be reimbursed his reasonable costs up to that date.

4    *Basis of negotiation*

- Agree a programme from the issue of the letter authorizing design and estimation, through the negotiation stage to the contract and on to completion of the job.
- Agree the general and special conditions to apply and any variables which affect the contractor's assessment of risk, e.g. substantial liquidated damages.
- Designs and specifications to be agreed to the maximum extent practicable, given the desirability of the earliest possible agreement on firm prices, between the contractor's and the purchaser's engineers/consultants. Commercial negotiations on price should follow technical agreement so that one is not trying to deal with the two variables at the same time. If during the course of the price negotiations it appears that the price is excessive for some technical reason then the issue can always be referred back to the engineers for further consideration and ultimately to the purchaser's management for an overall decision.
- The contractor should be instructed to prepare his estimates according to his normal method of estimating. When the estimates are complete (or largely so as it is not desirable to wait until the last few per cent if this will significantly delay agreeing firm prices on the remainder) then these should be gone through in detail and agreed with the purchaser. The methods of doing this are discussed in detail below.
- Once the estimates have been gone through and agreed the contractor should submit a normal firm price tender.
- After acceptance of the contractor's tender the contract should be treated in all respects as if it had resulted from a competitive bid, that is the contractor may gain or lose depending on how the job turns out and whether his assumptions and estimates were correct or not. There should be no reopening of the estimates, whichever way it may go for that particular contract. The purchaser may however wish to do so for repeat business. This is dealt with in more

detail below.

# Methods of price negotiation

There are basically two methods which can be used by the purchaser to negotiate the price. They are not necessarily mutually exclusive and indeed on a major project both should be used.

The first method is to compare the contractor's estimate for an item or section of the works with other prices already known to the purchaser or his consultant and which were obtained in competition for similar classes of work. At its simplest this method could involve comparing the square metre price proposed for a basic building with a price obtained recently by the purchaser's quantity surveyor on a competitive basis.

The main difficulty with this method is the obvious one of ensuring that one is comparing like with like and clearly the further the adjustments which require to be made to allow for differences the more spurious the comparison becomes. In the building example what services, water, drainage etc. were included, what are the standards of finish and fittings, are the construction conditions the same, etc. As the items being compared become more complex or the conditions of construction more divergent these difficulties increase. A one-off building constructed overseas in a developing country by a local contractor may differ widely in price from the same type of building constructed by an expatriate contractor as part of a total complex. The local firm is for instance unlikely to be over-concerned with compliance with safety standards in construction or the reasonableness of working hours or the accommodation/messing/recreational facilities for the staff/labour or with the final finish provided only that he can obtain a completion certificate. The UK firm on the other hand will apply its normal UK standards for safety etc., adopt a significantly different attitude towards its staff and labour, be concerned at the final results and be around to make good any defects.

On complex equipment, problems can arise on differing standards of design and specification, anticipated product life, environmental conditions under which the equipment is to work, degree of automation etc., all of which make useful comparisons extremely difficult.

The comparison method can only be used with any degree of confidence if:

1   On mechanical and electrical plant the specification and facilities provided are virtually identical. The most obvious example is where the same firm has tendered competitively elsewhere in the world for equivalent equipment. Even then allowance must be made for the commercial risk and risks applying to such other contracts in comparison with one now under negotiation; terms of payment, bonding arrangements and penalty clauses will all affect the price level. Then there is escalation since the base dates of the two contracts will be different.

2   On building and civil engineering work if the scope of work, specification construction conditions including programme and commercial terms are extremely similar. Within the UK this can be the case with UK standards and codes of practice and standard conditions of contract, and it can therefore make sense to do such comparisons at least for work above ground. Deep foundations, tunnelling or shaft sinking are another matter because of the great influence exerted on the pricing by the nature of the ground particularly the ingress of water and the presence of salts and acids.

   Overseas it is considered almost impossible to make comparisons which have real validity particularly between work executed in one country and another.

The alternative method is to require the tenderer to separate out his commercial overheads and profit and to break his unit costs and quantities for each item of work into its component elements.

On a building or civil engineering contract such a breakdown could consist of:

1 *Indirect preliminaries* This would be one sum for the contract covering general supervision, offices, camp costs, stores and plant yard common to the entire contract.

2 *Direct preliminaries* These would be associated with each section of a major contract covering the supervision up to foreman or sub-agent level for the section and any general facilities required for that section.

3 *Measured work* The labour, materials and plant utilized in the various operations, the quantities, times and rates of each being stated.

4 *Major materials and sub-contracts* There would be shown the prices to be paid and evidence that these were obtained competitively.

5 *Attendances and builders' work in connection* These would cover services provided by the main contractor to the sub-contractor, allowance for use by the sub-contractor of the main contractor's facilities etc., together with work by the main contractor in for example making good after the installation by the heating and ventilating sub-contractor of his pipework.

6 *Miscellaneous items* There are always a number of minor items included within any bill of quantities.

7 *Temporary works* The extent of these will vary considerably with the nature of the works in question. One important factor to note is the extent to which an item such as shuttering can be used a number of times.

8 *Design* If the contractor is responsible for design of the permanent works then this should be identified as a separate item.

9 *Contingency* This should again be separately identified and not hidden in the rates.

On mechanical and electrical contracts an appropriate breakdown would be:

1 Design charges.
2 Bought-in equipment.
3 Own manufacture identifying the shop costs for:

- materials
- labour
- overheads.
4   Delivery charges to site.
5   Installation establishment costs corresponding to the civil contractor's preliminaries.
6   Installation materials.
7   Installation supervision and labour.

In addition on both types of contract, if these are to be performed overseas the contractor should be asked to separate out any agency fees, although there may well be considerable opposition to this except on an extremely confidential basis.

## Discussion of costs and prices

Particular points which may arise in the price negotiations will include those described in the following paragraphs.

### Bought-in items and materials
The price of each should be checked against quotations or current estimates. It should be noted whether the prices quoted are fixed or variable, whether ex-works or delivered to site, and whether trade discounts have been deducted. Quantities should be checked to see that excessive allowances have not been made for wastage or contingencies.

### Labour costs in works
Wage rates and allowances should be checked against costing records. Overheads should be examined to ensure that appropriate rates for the contract in question have been used. For example, if a single works overhead is normally applied by the firm this may only be appropriate if the contract includes provision for a balanced workload of machining and assembly. If on the contract being negotiated machining is being sub-contracted and the contractor is carrying out assembly only, the overhead may require adjustment. If cost centres are used for

recovery of overheads, a check should be made to ensure that the ones appropriate to the class of work involved have been selected. Times for operations should be checked as far as practicable from the contractor's own records of past times for work of similar class utilizing similar method of production. If the contract involves substantial repetitive work, allowance should be made for the degree of 'learning' which will take place during the course of the contract.

## Design

The wage rates and overheads should be examined and a note should be made of the extent to which head office on-costs are being recovered through the drawing office. The best checks on design are to take the total man-hour quantities involved and see how these tie in with the programme, and secondly to compare the total allowance for design costs with the contract price as a whole. Experience will suggest to the negotiator the proportions of the contract price which should be represented by design. Another useful check is to take the quantity of drawings either produced or to be produced and arrive at a cost per drawing.

## Method statements

An important issue particularly on building and civil engineering contracts is the statement of the method by which the contractor intends to do the work – the combination of particular types of plant and labour. It is here that the contractor expects to make money from the use of his skill and initiative, but it is up to the negotiator to ensure that the method on which the costs are agreed is realistic and appropriate for the work in question. If later on during the execution of the contract the contractor can improve on it then that is his good fortune.

## Materials on site

The quantities of materials to be used will be checked by confirming the 'take-off' from the drawings with due allowance for wastage. Prices should be checked to see that they are competitive in relation to the quantities being used, and that discounts have been disclosed.

## Avoidance of overlaps

Particularly on large contracts, in which each section with its own supervision is estimated separately, the sum of that supervision and its related facilities will always be in excess of the total which the contractor will have on site. An estimator does not divide a man into two or even three, yet in practice one man will be found to be doing more than one job. Overlaps should always be looked for therefore as between sections and as between measured work and preliminaries.

## Labour

There are two elements to the basic labour cost: the rate paid per man and the man's productivity. Labour rates and associated benefits can be confirmed from the contractor's build-up of rates. Labour productivity is more difficult to assess except from experience of the particular work in question. One guide, particularly on plant installation work, is to look at the programme and the number of man-hours to be spent on site and see how these compare with the contractor's overall anticipated labour force. Again it is important to look at the picture over-all to avoid the problem of overlaps and to see whether the picture as a whole makes sense.

Overtime payments to labour should be identified and if the contractor is making any percentage charge on top of his labour costs for any elements of overhead recovery these should be related to basic costs only. It is preferable for overhead of preliminary costs to be assessed as items rather than percentages.

Charges for supervisory staff will normally include their benefits, such as company cars etc. It should be checked that these are not recovered elsewhere in the firm's overhead structure.

## Plant

Plant costs are of increasing importance and the negotiator needs to be assured that the basis on which plant has been charged is reasonable. In particular the following points arise:

1   On a large project, particularly overseas, certain plant should be capitalized.
2   The contractor will normally charge for plant which he owns at his own internal hire rates. It should be checked as to what elements (if any) included in these rates, e.g. profit, spares, servicing etc. are covered elsewhere either in his allowance for profit or in the preliminaries.
3   Is the plant being charged for the minimum time necessary? Negotiated contracts can often become the dumping ground for the contractor's own plant surplus to his immediate requirements.
4   Has the plant which is the most economic for the job been selected? This comes back to the method statement. One does not wish to find an expensive item necessarily used for a short period of time for a particular operation which then continues to be used on other operations for a much longer period of time, with intervals when it is standing, simply because it is then on site, when a much cheaper item could be used for those tasks.

# Head office charges

It is normal, and indeed desirable, that the firm should separate out its charges for commercial or head office overheads from the remainder of its costing structure. These overheads normally cover items such as the directors, company secretariat, research and development, legal department, central finance etc.

The list of items covered by the commercial or head office overheads should be examined to ensure that there is no duplication between these and any items which have been charged for in direct costs.

Two further points may arise in connection with the treatment of overheads. It is sometimes argued that if the contractor includes within his overhead build-up some item or service which is not required in connection with the particular contract under negotiation, such as expenses connected with export sales, then these should be deducted and the overheads adjusted accordingly. In principle this would seem quite wrong. In deciding to negotiate with a particular contractor, the purchaser is surely dealing with that contractor as a whole. He cannot select particular bits and pieces of the contractor's organization which have no separate commercial existence. Moreover, in fixing the overhead recovery rate the contractor will have taken into account the business which is generated by, for example, his export side and the contribution which that makes towards the general expenses of the business. The buyer cannot expect on the one hand to take credit for the turnover and on the other to refuse to contribute towards the costs which have made that turnover possible. The same reasoning applies to other services which the contractor maintains.

The second relates to the question of contingencies. Practice varies as between firms, but the most sensible way of dealing with contingencies would seem to be for the estimator to prepare his estimates as accurately as he can on the information available to him, and for the contingencies to be added as a whole to the total estimate by the sales manager or director responsible for deciding the final price level. If contingencies creep into the body of the estimate itself, as estimator's perks, then there is the danger of a double contingency being applied. It is not in the purchaser's interests to seek to reduce the final contingency below a sensible level. Any job carries unforeseen contingencies. If these are not allowed for initially they will form the subject of claims later on, and the lack of financial room within which to move may easily lead to delays on the job while extras are negotiated. The contingency must, however, be examined as a whole and considered in relation to the risks associated with the work and the profit which is being allowed.

## Terms of payment

An element of the contractor's pricing for the work will consist of his assessment of whether he will have a cash flow which is positive, neutral or negative. According to the nature of the business in which they are involved most firms will have included in their head office charges for the financing of the contract according to the normal terms of payment to which they are accustomed. Generally, even with building and civil engineering when payments are made monthly according to progress, this will involve some financing costs. With manufacturing, where payment is often delayed until delivery, there will be a more significant overhead charge.

The purchaser should establish the terms of payment which he proposes for the contract and then require the contractor to produce his estimated cash flow which is checked. Particular care needs to be taken over when the contractor is going to pay for materials and sub-contract work. A comparison between the terms of payment and the accepted cash flow will show the need, if any, to adjust the contractor's overheads.

## Price escalation

Price escalation is covered in detail in Chapter 20 and is still a vital factor in price negotiations even with today's more modest rates of inflation. Prices must only be agreed concurrently with agreement on whether they are fixed or variable and if the latter then with agreement on the escalation formula in the full detail on the indices, fixed elements, and method of application. It is only too easy for a negotiator to secure what he believes to be a favourable bargain on price from a contractor only to find later on that the contractor recovers more than he ever gave away as a result of the manner in which the escalation formula is applied.

## Equality of information

Where the contract is for an item which has been purchased previously, or the contract now being negotiated is for an item or service which will be wanted again, then from the purchaser's point of view it is desirable to establish if possible the principle of 'equality of information'.

All this means is that the purchaser is given reasonable access on a confidential basis to the contractor's manufacturing or other cost records, which are of course available to the firm's own estimators, so that both sets of negotiators start from the same point. There may well be reluctance on the contractor's part to supply this information, but without it the buyer is obviously at a disadvantage. If the buyer knows in advance that he is likely to be purchasing the item again on a negotiated basis, then he should seek to establish the position that he will be given 'equality of information' for the second negotiation, when he settles the terms for the first contract. It should be made clear that the information so provided for the second or subsequent negotiations will not be used to re-open the bargain for the earlier contract, even though it may show that the contractor has made a substantially higher or lower profit than was envisaged when the contract was negotiated.

## Proprietary equipment

Where the contract includes proprietary items of equipment manufactured by the contractor, the procedures outlined above for price negotiations are hardly appropriate and in any event would normally be unacceptable to the contractor. In this situation the fairest way of proceeding would be to require the contractor to satisfy the purchaser that the equipment being offered is competitive with that produced by other companies. Care must be taken to compare like with like and to make necessary allowances for differences in specification, performance, and capability. Also, if comparing list prices of equipment with prices included within a total contract, allowance must be made for commercial factors included in the latter for such items as overall management, penalty risks, financing terms, and so on.

# 6 Planning the tender

Since it is the purchaser who initiates the demand to which the contractor responds, the business of contracting has been looked at so far largely from the point of view of the purchaser. Having followed through from the planning of the project to the conversion of the plan into action by the issue of inquiries, it is now time to consider from the contractor's viewpoint the work and problems involved in tendering.

A tender is the most important piece of 'advertising copy' which a firm ever issues. Unlike most advertising material, it can be guaranteed that it will be read, and usually by the people who matter most. Not only, therefore, is it an important step in the chain of turning plans into physical action; it is also, for the contractor, a vital opportunity to project himself and his products, not just for the particular job in question but for the future as well.

There is much more, therefore, to tendering than the mere setting down of the specification, prices and terms on which the offer is made. There is the psychology to be studied of the buyer who will receive the bid; the importance to be examined of this tender in relation to the market as a whole and to the totality of the contractor's business with the customer concerned; the likely actions of competitors to be considered, and so on (see the author's *The Art of Tendering*).

Before putting a tender together, therefore, the contractor will normally take the following action:

1   Make a careful study of the inquiry documents.
2   Based on that study and on the information gained through normal commercial intelligence channels, and taking into account his existing and projected work load, decide whether to treat the inquiry seriously or not.
3   If the decision is to take it seriously, then prepare a tender plan, since a contractor must plan his tender in the same way as a purchaser must plan his project.

## Study of the inquiry documents

The type and character of inquiry documents vary tremendously. On the one hand there is the simple letter asking for a quotation to be submitted; on the other the massive commercial/technical documents issued by large customers and consulting engineers, often with specific tender forms which the tenderers are required to complete. Certain problems are common to both and to the wide range of documentation in between. A checklist of commercial questions, including those which would be relevant if the works are overseas, which should be answered before the decision is taken to bid is given below:

1   For what work is the contractor to be responsible? Are the terminal points clearly defined?
2   Is it clear what the employer is going to provide or do and by when? Who is responsible for the interface between the contractor's and the employer's work? Are the employer's obligations stated in such a way that they are contractually binding on him? What is the risk of his defaulting on these?
3   Does any part of the work involve:
    ● adaptive engineering

- development
- use of non-proven components or techniques?

If so what is the extent, how near is it to the 'state of the art' and what would be the consequences of failure?

4 Does the contract clearly define in relation both to factory testing and site testing:

- the type and specification of tests to be carried out
- test limits
- objective standard for visual tests
- procedure for repeat tests
- when and within what period tests are to be carried out
- that no additional tests can be added by the employer beyond those specified in the contract
- whether the employer will repeat tests or observe the contractor's tests?

5 Are there guarantees for performance and penalties for failure to meet these?
   If so then:

- when will the guarantee tests be carried out?
- who will operate the plant during the guarantee tests?
- who will provide the necessary facilities for the carrying out of the tests?
- who provides the test equipment?
- are the limits, tolerances and test methods specified?
- what happens if the employer is unable to have the tests carried out when the contractor is ready? Is the contractor then entitled to have the plant taken over? Does the contractor have to carry out the guarantee tests then during the defects liability period?
- is there provision for a reliability run? If so, when does this take place, what are the conditions for the turn and in particular what are the permitted outages?

6 To whom will the contractor be responsible – directly to the

employer or to another contractor? What is the financial standing of the employer or main contractor? Will there be an engineer under the terms of the contract and if so who will exercise his powers?

7    What are the contractor's obligations in relation to time for completion? Is the contract programme a contractual document so as to make the contractor contractually liable for meeting intermediate dates? Is completion itself clearly defined and is it before or after the performance guarantee tests or the reliability run? Is there an escape clause if the works are substantially completed? Are there liquidated damages or penalities for delay and if so at what rate are these, is there a maximum and what are the contractor's liabilities if the maximum time limit is reached? Can the employer terminate for delay or claim consequential damages?

8    What are the general conditions of contract? Are there special conditions added and if so what are these? Do the conditions of contract and the specification contradict each other? Do the conditions impose any special risks in relation to the nature of the work to be carried out?

9    What are the terms of payment proposed and would these produce a negative or positive cash flow? What bonds is the contractor required to provide and are these cashable on first demand? If a bond is cashed must it be replaced? Is there any requirement for credit finance? In what currency and where will payment be made? Is there a risk on exchange rates?

10   What are the contractor's responsibilities in relation to insurance? Is he required to insure with an overseas insurance company and if so in what currency will payment be made and what is that company's record on claims payment?

11   What are the contractors' liabilities in relation to defects? Is the defects liability period revolving? Is there any liability for consequential damages?

12   Does the contract allow for extensions of time and if so for what reasons? What is the procedure for claiming extensions and how are these assessed?

13 Are there any nominated sub-contractors proposed? If so are they commercially acceptable and is it necessary to contract out of any risks in relation to them?

14 Under what legal system will the contract be governed and how will disputes be decided? In what country are the assets of the employer/main contractor situate?

15 What are the employer's rights to terminate and what are the consequences of termination?

16 How much time is available for tendering? Are there any special formalities attached to tendering such as submission in a foreign language or notarized copies of the tender?

17 By whom may the tender be submitted? Are there any rules governing the employment of agents?

18 What is the contractor's liability for the payment of overseas taxes either in respect of profits or on the salaries of his staff? Is there any double taxation relief between the UK and the country where the contract is to be performed? Are there any special requirements on import permits, visas or work permits? Are there any special fees or taxes payable on imported materials and plant? Are there stamp duties payable on the contract and if so by whom?

19 Is the contract fixed price or subject to escalation? If the latter, how is escalation to be calculated? Are there reliable statistics or indices available in the overseas territory?

20 Is the final certificate issued at the end of the defects liability period conclusive evidence of the sufficiency of the works, or does the contractor have a continuing liability? If so, for how long?

It is to be hoped that the answers to the more general of the above questions, such as those relating to law and taxation, and indeed to those relating to the employer, are already known to the contractor from his previous investigations of the market. If they are not, and he is starting in a new territory from scratch, then, as suggested in *The Art of Tendering*, p. 38, the contractor is almost certainly wasting his time and money in preparing a bid.

## Planning the tender

The tenderer's objective is the submission of an offer which:

● is the most attractive to the customer which can reasonably be presented
● minimizes the contractor's risks and potential liabilities and ensures the contractor a reasonable profit return.

Clearly these two objectives will at times be in conflict with each other. Thus it may be attractive to the customer to guarantee a twelve-month delivery when one's competitors are only willing to offer eighteen months, but if the damages for delay are 0.5 per cent of the contract price per day, the tenderer must be very certain of his ability to complete on time for the risk involved to be commercially acceptable.

Thus tendering, like purchasing, is a compromise. Moreover, it is a compromise which normally has to be worked out against a tight time scale and, unlike purchasing, has to take into account the activities of the firm's competitors. It also costs time and money and is a commitment on a company's resources. Planning may, therefore, be considered in two stages: first the decision whether to tender at all, and second, if the decision is to go ahead, the planning of the tender itself. The first issue, that of the bid/no bid decision is covered in some detail in the author's *Contract Negotiation Handbook*, pp. 48–67. The following is a summary of principles outlined there and of the manner in which it is suggested that the contractor should approach the matter.

There are two factors involved in the bid/no bid decision: bid desirability and success probability. It is suggested that the firm should initially analyse the invitation to tender in terms of its bid desirability using for this purpose the questionnaire and marking scheme set out in Appendix 2 which is reproduced from pp. 53 and 54 of the *Contract Negotiation Handbook*, and mark the answers as there suggested. If it is apparent that any factor which cannot be changed is strongly negative – such as a mandatory requirement to accept payment in a non-freely convertible foreign currency – then the decision should be no bid and this is so regardless of success probability. It is important at this stage to be totally realistic in recognizing those factors which are mandatory and will not be changed by the employer, otherwise a bid may be submitted with qualifications and bid bond lodged, and then the firm be advised that it has been awarded the contract and instructed to come and sign the contract on *the employer's terms and with the contractor's qualifications deleted*. It will be useless at this stage to protest or prevaricate. The firm will have only the option of signing on the employer's terms, with all the risks these involve, or of forfeiting their bid bond and suffering the financial loss.

Assuming that the bid desirability appear reasonable then the firm should estimate its probability of success related to that of the other firms which it believes are likely to tender. For this purpose the firm should rank itself against the other potential tenderers looking so far as it can through the eyes of those in the employer's organization who will share in the buying decision. For the moment the firm should assume that it will tender at its normal level of pricing and on the basis of only accepting those contractual risks which it normally accepts in relation to the class of work in question.

The method used for assessing success probability is in outline as follows, the firm being referred to as party. For a more detailed account see pp. 67–69 of the *Contract Negotiation Handbook*.

| | |
|---|---|
| *Step 1* | Party selects from the list set out in Appendix 3 the factors which it considers will be most relevant to the employer in comparing party with its competitors. |
| *Step 2* | Each factor is allocated the same number of points and party ranks himself against his competitors. |
| *Step 3* | Party identifies the functions within the purchaser's organization which will contribute significantly to the buying decision. |
| *Step 4* | The value of each factor assessed under Step 2 is weighted by a combination of: |

- the value each function in the purchaser's organization will give to that factor and
- the influence level which that function possesses in the making of the buying decision.

It may be suggested that any such exercise must necessarily contain a significant margin of error, and that the figures will tend to be adjusted to give the answer that people want. The reply to those criticisms is first that the exercise should always be done independently by at least two people from different departments, and second that the simple requirement of making people give consistent quantitative answers to questions imposes its own discipline and makes people think more clearly. It is however agreed that the results must be interpreted broadly.

Having now arrived at a figure for success probability this can then be combined with that already obtained for bid desirability to give an expected value for function for the bid. If this is above the minimum level established by the firm at that time for that class of business, and no better opportunity exists, then party should bid.

Assuming that the decision is to bid then the firm should take the following actions in order both to maximize its chances of success and minimize the risks should it be successful:

1  Appoint a tendering team with a tender manager.
2  Ensure that it has the appropriate 'political' representation necessary to support its interests. Again this should already be in place if the firm is to have a real chance of winning – see further Chapter 8, *The Art of Tendering*. Now is the time to ensure that that representation is actively at work.
3  Visit the site armed with a questionnaire to complete – see Appendix 4 for a specimen.
4  Identify from the bid desirability table any particular actions which can be taken to minimize risk or improve success probability, allocate responsibilities for these to individuals, and follow up and assess the results achieved.
5  Seek clarification from the purchaser even if only informally on any ambiguities in the tendering documents which unless

resolved would make it necessary to include reservations in the tender.

6 Obtain specific local advice on any matters of law, taxation, import regulations etc. which could affect either risk or price or both.

7 Establish whether or not the purchaser would be receptive to any alternative, either technical or commercial, which would increase the firms' success probability. Further whether he would be prepared to award the contract on the basis of an alternative, either without giving the other bidders an opportunity to re-tender, or only a nominal one.

# Tender price level

The firm's tender price level will be a function of the following factors:

1 The buying policy of the purchaser. Does he negotiate with the low bidder or the lowest two in order to secure reductions in the tendered prices or not? If he does then the firm must allow a margin above their minimum price level in order to be able to satisfy the purchaser's requirements. If the firm is uncertain as to the purchaser's policies then for his own security he should assume that the purchaser will negotiate.

2 The worth at any given price level which the bid would possess for the purchaser. This brings into account the non-price factors such as delivery, technical merit, proven record of performance etc.

3 The anticipated bidding strategy of the firm's competitors.

4 The worth to the firm of a bid at any given price level. This brings into account the state of the firm's order book, current level of activity, future marketing policy, contractual risks associated with the contract, financial considerations such as cash flow, bonding requirements etc.

Based on these factors it is proposed that the firm's decision rules on bidding can be summarized as follows:

1   *Competitive bid – purchaser not expected to negotiate on price.* Bid at the level which will maximize the bid's subjective expected value to the bidder, i.e. the product of the success probability of a bid at that level and its worth at that level to the bidder.
2   *Competitive bid – purchaser expected to negotiate on price or bidder uncertain as to purchaser's intentions.* Bid at the level which is marginally above the assumed worth of the anticipated lowest competitive bid to the purchaser, after taking into account non-price factors provided that it is above the bidder's minimum acceptable price level, i.e. includes a margin for negotiation. This will ensure that the bidder is invited to negotiate and does not reduce the upper level from which the bargaining will commence.
3   *Non-competitive bid.* Bid at the level at which it is believed that the purchaser would just be indifferent between placing the contract and not doing so, adjusted to take account of the time-costs associated with achieving agreement having started the negotiations at that level, provided again that this level is above that which would be the minimum acceptable to the bidder.

For a more detailed treatment of the above together with worked mathematical examples see Chapters 5 and 6 of the author's *Contract Negotiation Handbook.*

# 7 Joint ventures and consortia

Joint ventures may be entered into for a variety of reasons some of which may be termed aggressive in that they seek to bring together a combination of skills which is best able to undertake the work on a turnkey or main contractor basis. Others are defensive of which the most common is quite simply to reduce the competition. Or the joint venture may be a 'shot-gun marriage' in that in many territories today – unless the job is being funded by an international lending agency – there is simply no way in which a foreign contractor can be awarded a government contract up at least to a certain value unless he has a joint venture with a local partner.

All joint ventures for whatever reason they are undertaken share certain characteristics and have certain problems which must be solved *at the outset* or else the relationship has a high probability of ending in disaster. Joint ventures with local partners overseas additionally present certain difficulties of their own which are discussed later in the chapter.

## Joint venture characteristics

The terms joint venture and consortium are often used loosely without proper definition. Here joint venture will be used to describe a relationship in which the parties have agreed to undertake the contract on an integrated basis in which each provides staff and resources which are combined together, and no one party is separately responsible for any individual section. In a consortium in contrast each party is wholly responsible within the consortium for the pricing and execution of a particular session of the work. The internal arrangements do not normally affect the employer since he will insist that the parties – whether it is a joint venture or a consortium – are jointly and severally liable to him for the performance of the contract as a whole.

The distinction has an important impact on the internal structuring. If it is a consortium and not a joint venture then there will be a need for cross-indemnities between the parties so that if one party fails to perform and the others have to fulfil his obligations, then they are protected against the consequences. This is, in practice, easier said than done since what is required to be assured is the financial worth of the party in relation to the obligations he has undertaken to perform. This may need the support of on-demand bank guarantees which, if they are not forthcoming, are a fair indication of the value to be placed on the indemnities. Also if the work performed by the member who has defaulted is of a highly specialized nature it may be difficult to find a replacement.

The sharing of profits or losses as between the parties is also significantly affected by the decision on the form of co-operation. If it is a joint venture then this will normally be pro-rata to the value of participation and profit will usually only be taken at the joint venture level. If however it is a consortium then each party will take the profit or loss on his own work and it is then necessary to decide how to handle the consortium costs. Often the decision will depend on the local rules as to taxation and tax advice on this issue should always be obtained before any decision is made. What must be avoided is so-called 'cascade taxation' in which profits – or what is worse, deemed profits – are taxed at both levels.

Another issue to be determined is as to whether the joint venture or consortium should be incorporated or not. Incorporation often has advantages structurally and may in certain territories be a political, if not a legal, requirement. However it can have distinct tax disadvantages, one of which is that assuming the company is being incorporated overseas, the UK parents will not be able to claim tax relief on their marketing expenses. Again tax advice both at home and abroad must be obtained before any decision is taken.

The key issues which should be covered in the Joint Venture Agreement, apart from those already discussed, are as follows:

1  *The objective.* Is it pre-bid only, to bid for a particular project or is a longer-term relationship envisaged?
2  *The duration.*
3  *The law of the agreement.*
4  *Procedure for settlement of disputes.*
5  *How is the agreement to be managed?* There are several issues here which require to be considered:

  ● Is one company going to act as the sponsor? If so the responsibilities of the sponsor need careful definition, particularly as to the limits to which he is entitled to commit other parties. The sponsor's fee must also be settled. The advantage of a sponsor, particularly operating overseas, is that it enabled the overall management to be handled through an existing organization and one which has already established links with the agent.

  ● A management board needs to be established which is comprised of senior members of the parties who have sufficient time and a sufficient degree of availability to attend to the business. Again when operating overseas the question of availability is extremely important. It's no use appointing people who are unable to attend meetings because of other commitments. The terms of reference of the board must be defined. This raises the issue of what constitutes a quorum and voting rights which may appear matters of detail but can become extremely important when there are issues of great financial import on the agenda.

- A project director has to be appointed to exercise day-to-day managerial control reporting to the management board. This is a key role the essence of which is *management*. If the project is overseas he must have a good up-to-date knowledge of the territory and how business is conducted there and be personally acceptable to all locals who may be involved.

6  How is the tender price to be built up? Is the pricing of particular types or sections of the work to be done by one party or by two separately and then estimates compared? Policies must be agreed upon for the handling or risk and contingencies.

7  The approach to the tender conditions and qualifications must be settled. Usually it pays to appoint the party having the best experience of dealing with the particular client to handle this issue and prepare proposals for ratification by the management board.

8  The procedure for contract negotiation with the employer needs to be determined. How is the negotiating team to be constituted and what authority will they possess? Do all possible changes to the tender have to be referred back for unanimous agreement? This may be desirable but is it realistic? If not, how is the problem to be handled?

9  Confidentiality of information provided by one party to the others must be covered. Also non-disclosure outside the joint venture other than for the purposes of the joint venture.

10  It is usual for the parties to agree to participate on an exclusive basis and this can be very important where one party may be approached by a competitor to act as a sub-contractor.

11    Financial considerations will include the following:

- The establishment of a budget for the tender and the apportionment of tendering costs. Alternatives are that each party pays his own costs for the services which he contributes and then certain common costs are shared pro rata to participation, or that all costs are pooled on an agreed basis and then paid pro rata. In this latter event there must be provision for independent auditing.

- How are the parties going to share in the provision of the bonds required by the tender? Although the bonds for the benefit of the employer will have to be joint and several it can be possible to arrange the recourse to the bank issuing the bonds on a several basis pro rata to each party's portion of the work where the work is being executed not on an integrated basis. Alternatively a bank appointed by the joint venture can be asked to package and charge each member company on a joint and several basis. This will mean a higher charge for some than others because the bank will probably not assess each firm on the same basis but can produce overall savings.

- If a financing offer is required than a financial adviser, usually a merchant bank, will need to be appointed.

- The accounting arrangements covering the receipt of funds from the employer, their employment and their distribution must be defined in some detail. If at least part of the payments are in a foreign currency then management of the exchange risk will be important.

12    The retirement or possible expulsion of one party from the joint venture should be covered, together with his continuing obligations on confidentiality and non-competition. It is usual to provide that a party can withdraw up to the time of submission of the tender but not thereafter unless all other parties agree. With a consortium as opposed to a joint venture the retirement of one party may make completion of the contract work difficult. Account must also be taken of the provision on this point in the contract with

the employer.

## Special considerations applying to local partners

The first point to establish is why a local partner is being included. Possible reasons are:

- because it is required by local law or practice
- to gain a political advantage because of his connections with the employer or others involved in the contract award
- because of his knowledge of local working conditions and ways of doing business
- to reduce the tender price
- to allow part of the price to be tendered in local currency where this is not freely convertible.

In practice, more than one of these reasons may apply but the essential point is to distinguish between a local partner who is essentially included for his connections, and one who is intended to participate actively in the execution of the contract work. In the former case the local will have to be 'carried' by the foreign partners and it will not be practical to expect him to assume genuine responsibilities for work performance, the provision of bonds etc. Equally he cannot expect to have any genuine say in the way in which the contract is managed and performed and he will have to be content with a reduced level of profit or even simply a fee.

In the latter case he has to take a share in the project risks, performance and rewards or losses so far as he is able financially to do so. The proviso is important since many potential local partners. overseas are undercapitalized and with a very thin layer of competent management. The other point to appreciate in advance is that their methods of estimating and work management/execution and attitudes towards contract conditions and risk may differ significantly from those to which the foreign partners are accustomed. These issues need to be discussed frankly but sympathetically and without the degree of arrogance which only too often foreign partners display on these occasions. Their resolution must not be left to the stage of tender finalization.

# 8 Tender preparation

In the actual drafting of the tender the contractor has to satisfy as far as he can two conflicting objectives. On the one hand the primary function of a tender is to act as an aid to selling. Through its medium the contractor is seeking to persuade the buyer that he, rather than any other, should be selected for the award of the contract. Its preparation should therefore be attractive and positive. At the same time the tender is the contractor's opportunity, often his only opportunity, of seeking to protect himself against provisions in the inquiry which he considers are unreasonable. At the least, if there are any such provisions, he must make certain his tender is so worded that it cannot be accepted without his having the right of discussing these with the buyer.

Regarded as a 'selling' document, the most important points to be considered in drafting the tender are:

1   Meeting the purchaser's essential requirements. If, for example, the purchaser's prime interest is in having a price within a week, then he must be given the price within a week if this is humanly possible, if necessary by telex. The

technical and commercial details can follow.

2 Demonstrating to the purchaser the skill and efficiency of the contractor. A purchaser may well consider that a 'sloppy' tender is evidence that the job will be carried out in the same way. Therefore, within the limits set by item 1 above, the tender should be well presented, clearly readable, indexed, if of any length, and should hang together as a whole. It should not, for example, contain copies of sub-contractors' quotations with their terms of sale attached, which are nothing to do with the purchaser.

3 Bringing to the purchaser's attention those points which, judging from the inquiry, are those in which the purchaser is most interested and where the tenderer can stress the technical or other advantages which he believes his offer has over those of his competitors. It is no use expecting the buyer to guess at these, and it is equally dangerous to assume that he will delve deeply enough in his tender appraisal to establish the true value of one offer as against another. He may, if he has the time and the ability. Far better to present the information to him in such a way that he cannot overlook it. It is rather as if the buyer were an examiner and the tenderer the pupil. The buyer is no more entitled to make assumptions than the examiner is entitled to guess at his pupil's knowledge of the subject. Both can judge only on the data presented to them.

Looked at the other way round, as a 'protection' document:

1 If there is any item over which a doubt could arise as to whether it is included or not, then the tender should make this clear. If, for example, in an installation contract the tenderer is not including an allowance for lifting tackle for off-loading purposes, then he should state this specifically. There must always be a statement defining the limit of supply and a schedule of specific exclusions.

2 If the inquiry includes terms and conditions which the tenderer considers unreasonable, it is often difficult for him to decide what comments to include in his offer. Some forms

of inquiry either include statements to the effect that any qualifications made by the tenderer may lead to his being disqualified, or require that the tenderer should give specific confirmation in his tender that he accepts the terms and conditions offered. In any event, a long list of suggested modifications to his proposed conditions of contract may lead to the buyer becoming suspicious or impatient with the tenderer, and so to the latter losing the order. On the other hand, terms which could easily involve the supplier in heavy additional expense may create risks which the contractor considers are unacceptable, having regard to the price level of the contract. In those circumstances he must as a minimum make clear in his bid that he has certain objections to the terms proposed and would wish to discuss these if his offer is otherwise of interest. This at least establishes his right to negotiate. If the purchaser's terms are inappropriate – if, for example, they are 'supply-only' conditions for a job including erection and commissioning – then the tenderer could state he has no objection to the purchaser's terms as such, but would propose Form . . . which he considers more suitable for this particular contract and on which he has based his tender. He would be happy to discuss and agree with the purchaser on the conditions to apply to the order.

## Tender documents

Leaving aside those cases where the buyer issues his own form of tender for the supplier to complete, a tender for plant or equipment will generally consists of:

- the covering letter
- the specification
- list of exclusions and schedule of services to be provided by the purchaser
- terms and conditions of sale
- the quotation.

One other document which the tenderer may wish to prepare in particular circumstances is an executive summary of his tender. The decision-making process for large engineering works will usually involve those at the very top of the purchaser's organization and at times, especially overseas, those who have a political interest in the project extending occasionally to the Prime Minister of the country itself – see Chapter 8 of *The Art of Tendering*.

In this type of situation the tenderer either through his own contacts, or more likely overseas, his agent, should take steps to bring the key features of his proposal to the attention of those with political power over the decision and whom he believes can be influenced in his favour. Such people are far too busy, apart from not being appropriately qualified, to read pages of detail. If the tenderer does not take the initiative himself his agent – if he is any good – is sure to ask for an executive summary and the experienced tenderer should have one ready.

What the summary should contain will obviously vary from case to case but as a guide the following points should be covered in an overseas situation and many of them with perhaps a slight difference of emphasis will apply domestically:

1   The basic contract price. Optional extras such as training and spares should be omitted.
2   The completion period.
3   The main financing terms presented in as positive manner as possible.
4   Benefits which acceptance of the offer will provide to the country with emphasis if possible on the part of the country in which the politician is known to have a particular interest. Such benefits would include:

    ● transfer of technology
    ● use of local manpower and material resources including local consultants/suppliers. Any firm in which the politician is known to be personally interested should certainly be mentioned.
    ● savings on foreign exchange due to the ability to take

payment in local currency.

5   If it can be said (see again Chapter 8 of *The Art of Tendering*) that there is British Government support for the bid. A letter confirming this from the local ambassador is always useful.
6   The long-term interest in, and commitment of the tenderer to, the country concerned as evidenced by the formation (if this is the case) of a joint venture company with a local partner, or at least his having established a permanent presence there.

## Covering letter

The aim should be to keep this as short as possible. Ideally a covering letter should do no more than:

- introduce the tender and identify the documents of which it is comprised
- state if any alternative scheme or proposal is being submitted and where this can be found in the tender
- refer briefly to any particularly important aspect of the offer and whereabouts this is set in more detail
- if there are any major reservations on the terms of the inquiry, refer to these.

One reason for avoiding detail in the covering letter is that after initial study it may become detached from the tender itself and be placed on a correspondence file, and so not be referred to subsequently during the tender appraisal. Nor is there any point in duplicating in the covering letter information which is already contained in the tender. Further, because there may be doubt as to whether the covering letter forms part of the contractor's offer in the contractual sense, the covering letter should not be relied upon to establish contractual rights. These should always be set out in the body of the tender itself. For an example see the celebrated case of *Davis Contractors Ltd v Fareham UDC* [1956] AC 696. Here the contractor's covering letter to his tender which did contain reservations on his absolute responsibilities for the supply of labour and materials was not referred to when the contract was placed and so did not become part of the contract.

As a result the contractor was held liable to complete his contract without any right to claim any extension of time for delays due to labour shortages.

An example of a covering letter for a tender for major plant and equipment might be:

With reference to your inquiry number 4563/68 dated 30 June 1995, we have pleasure in enclosing our tender in two volumes, lettered *A* and *B*, together with a separate folder of drawings. Volume *A* contains our offer together with a general description of the plant. Volume *B* contains our detailed specification.

We have put forward an alternative layout for the coal handling section of the plant which we believe will provide substantial economies both in capital and operating costs. Full technical details of this alternative are given in volume *B* section 2, and the price reduction we are able to offer is shown in page 21 of volume *A*.

In view of your interest in the plant being operated with the minimum of manpower we would draw your attention to the comprehensive remote monitoring and control scheme described in section 1 of volume *A* and to our substantial experience in this field, full particulars of which we have set out in that section.

We have carefully considered the Terms and Conditions subject to which your inquiry was issued. In general we think these to be very fair and reasonable, but there are just one or two reservations to which we have referred in section 4 of volume *A*, which we would like to discuss with you in the event of our tender being of interest.

We hope that you will find our proposals satisfactory and we shall be pleased to give you any information which you may require.

### Specification

This really falls into two parts: first, the general description of the plant being offered, and second, the detailed technical data.

In terms of layout it is suggested that the tender should start with a general description of the plant written in such a way that it is interesting to read and can be understood by the customer's senior management. This is the tenderer's 'shop window'. This, plus the actual offer, is probably the only part which the customer's senior management will read. It should therefore be made comprehensive and stress all the main technical features and advantages which the tender contains, but without obvious sales 'padding' and avoiding the use of sales jargon. It is also the place where the tenderer can stress his previous experience in the field to which the tender relates. It is only too easy for a firm to assume that the customer, because he has put him on the tender list, is aware of the work which he has done. In fact this is often not so, and it is always worth while for a firm to educate the buyer in this respect.

This part of the specification can conveniently contain a summary of the main equipment offered together with a list of the terminal points and exclusions. This will be extremely useful for the customer's purchasing and managerial staff in comparing the broad extent of supply of one tenderer against another.

The preparation of the detailed technical specification will obviously vary tremendously with the type of plant being offered, but some suggested points for consideration are as follows:

1  Make it easy to read and follow. Remember that the customer's engineers have only a limited time in which to study the offers.
2  If the customer has not indicated how he wants the specification sectionalized then there are usually two possibilities. In the first case the tenderer should give complete physical sections of the plant including all types of equipment within the section. This can often be conveniently related to the sectional breakdown of prices called for in the price schedule. Alternatively the tenderer can specify type of equipment or processing unit. Thus all the mechanical equipment might be in one section, the electrical equipment in another, and the civils and structures in a third. This can be convenient in that the customer's engineers need only then read that section which concerns them. Whichever way it is done a comprehensive index is required.
3  If the customer has provided schedules for the tenderers to fill in, these should be completed in accordance with the

customer's instructions. The tenderer should never attempt to know better than the customer how he wants the bid presented.

4 Ensure that information obtained from sub-contractors and suppliers is properly integrated into the tender. Cut out from their quotations material which is irrelevant as far as the customer is concerned, and make sure that the whole document reads as one.

5 Use common item numbers throughout the specification and drawings for easy identification.

6 Make the maximum use of schedules for giving technical data and characteristics of equipment being offered, for example motor and pump schedules as opposed to pages of description which are tedious to read.

### Alternatives

One point which sometimes arises is whether or not to include an alternative design which may be cheaper or possess some technical advantage over that on which the customer has required that the main offer be based. The problem in disclosing the alternative at tender stage is that, once it has been submitted, the buyer may take the view either that he must obtain competitive quotations for the alternative from other tenderers, or that at least he should give the other tenderers the opportunity of submitting their own alternative proposals. In either event the firm may lose the commercial reward which their ingenuity should have earned for them. Much depends on the tenderer's view of the action which he considers the buyer is likely to take. If he can be reasonably confident of getting a fair deal, then he is probably best advised to disclose the alternative in his tender, so as to be sure that it is taken into account when the tender comparison is made.

### Exclusions and services to be provided by the purchaser

Ideally this schedule needs to be sufficiently comprehensive to prevent any doubts arising later as to whether a particular item was included in the offer or not, or as to the extent of the services which the purchaser is required to perform. In the initial stage of submitting his tender, however, the tenderer may decide to leave himself room to negotiate and not be entirely specific, so that, when called to discuss his offer, he can play it according to his judgement of how his bid stands in relation to those of his competitors. These can be legitimate tendering tactics, but there is clearly the risk of being caught and of either having to provide more service to the purchaser or receive less from him than was envisaged when the tender prices were prepared.

## Terms and conditions of sale

If the purchaser has not stated any terms or conditions of contract in his inquiry, then it is open to the contractor to submit his offer subject either to his own individual terms or in accordance with one of the standard sets of conditions of contract published by the engineering institutes or the contractor's own trade association. Generally it is in the tenderer's interests to satisfy the purchaser that he has taken an objective attitude in respect of terms of contract, and from this point of view it is easier for him to do this by using a standard institute form than by using one which he has prepared himself. The latter is bound to be looked at by the purchaser with some suspicion. Moreover, the purchaser's staff will probably be familiar with the institute form, and thus the tenderer will again earn favourable marks by having simplified and reduced the work of tender appraisal.

The same situation arises in reverse when the purchaser states the terms and conditions in his inquiry. If these are one of the institute forms, perhaps with minor modifications to suit the purchaser's particular circumstances, then the contractor can normally accept these without any difficulty. If, however, the purchaser has prepared his own conditions, then the contractor is bound to regard these as being subjective in their approach and to submit them to a critical examination. If the result of such examination is that the contractor considers the conditions are more onerous than he is prepared to accept, he is often in something of a dilemma as to the extent to which he should make his objections known in his tender. On the one hand he does not wish to offend the purchaser or in an extreme case disqualify his bid from being considered; on the other hand, unless he makes some reservations at tender stage he may be taken to have accepted the conditions without qualification.

If the contractor does consider the terms offered unacceptable, then as a minimum he must make it clear that there are certain points which he would wish to discuss in the event of his tender being otherwise acceptable. How much further he goes in being specific as to his objections or in putting forward alternative conditions of contract must depend on the circumstances of the particular inquiry, and the view which the tenderer takes as to the purchaser's likely reactions. In making his decision the contractor should take into account the following points:

1   If the terms offered are wholly or largely unacceptable, then the tenderer must put forward an alternative basis, and this should be as objective as possible.
2   If the purchaser is likely to place the order without post-tender negotiation, then again the tenderer needs to submit his offer in a form in which it could be accepted; that is, if there are particular clauses to which he objects he should propose alternative drafting.
3   If on the other hand there is likely to be room for negotiation, then the tenderer may be in a better position if he merely indicates his objections in principle but without drafting.
4   The character of the purchaser's staff and their degree of sophistication in commercial matters.
5   Any known rules or procedures established within the purchaser's organization, for instance that modifications to standard conditions have to be submitted to head office, which is almost always unpopular. Here it may be possible for the tenderer in negotiation to achieve the desired result in some other way, for example by a side letter to the contract. In his tender, therefore, all he would be advised to do would be to establish a negotiating position.

## Quotation

There can obviously be no standard form for this, but there are a number of points which normally require to be considered as follows:

1  The validity of the offer. Although a promise to keep an offer open for a certain period is not legally binding, unless the purchaser has given consideration for the promise, it is important commercially for the tenderer to make clear the validity period of his offer. This gives him the opportunity of revising his offer once its validity has expired without being accused of acting in bad faith.

2  Whether prices are fixed or subject to price escalation. If the latter, the basis on which price escalation is to be calculated.

3  Whether the individual prices in a schedule of prices constitute separate offers, or whether the only price which is open for acceptance is the total for the schedule.

4  If fees are quoted as a percentage it must be made absolutely clear what is the base to which the percentage is to be applied.

5  If a rebate or discount is payable above a certain minimum figure, whether this is calculated on the whole of the sum or only on that part which is in excess of the minimum. An example may make this clear. On a tender for the hire of constructional plant the tenderer offers a deferred rebate according to the value of plant hired from him during the year according to the following scale:

$$
\begin{array}{ll}
\text{Over £100 000} & 2\frac{1}{2}\% \\
\text{Over £150 000} & 5\% \\
\text{Over £200 000} & 7\frac{1}{2}\% \\
\end{array}
$$

If the total value of plant hired is £230 000, this is capable of two interpretations:

- that the whole £230 000 is subject to a discount of $7\frac{1}{2}$ per cent, that is £17 250
- that only the excess at each stage is subject to the appropriate rate of discount, that is:

$$\text{£50 000 at } 2\frac{1}{2}\% \ = \ \text{£1250}$$

$$£50\ 000 \text{ at } 5\% \quad = \quad £2500$$
$$£30\ 000 \text{ at } 7^1\!/_2\% \quad = \quad £2250$$

| | |
|---|---|
| Total | £6000 |

It hardly needs to be stressed how important it is that the offer is written in such a way that there is no ambiguity as to what is intended.

6   If any item is described as provisional, but is later to be converted to a firm price, that there is included somewhere within the terms of contract a statement as to how this is to be done and what factors are to be taken into account. Is the contractor, for example, entitled to make adjustments in his basis of pricing because of events which have happened or knowledge which he has gained subsequently as to the conditions under which the work will be executed?

7   If the value of any item is to be determined according to the quantity of work done or services provided, that again the mechanism for doing this is clearly established. If a budget estimate for such items is given it must be clear whether this sets a contractual ceiling or not.

8   If any work is to be executed on a daywork basis, then the items included within the percentage on-cost, the base to which the on-cost is to be applied, and the hours for which payment is to be made need to be clearly set out. For example, is the percentage applied to the actual wages paid, including bonus and/or overtime? What grade of supervision is within the on-cost percentage? Is travelling time to be paid for by the purchaser?

9   Is all overtime included within the contract price, or is overtime over a certain limit to be paid for, and if so on what basis?

10   In respect of imported items, are freight, import duty and the like included, and who is responsible for fluctuations in the rate of exchange if any payments are to be made in foreign currency? See also 'Sufficiency of tender' on p. 211.

# 9 Tender appraisal

The tenders having been prepared and submitted, the purchaser now has the task of tender analysis. There will be considered first the appraisal of tenders for plant and equipment and process plant and then tenders for civil and building works. The appraisal of offers submitted by competing tenderers for plant and equipment or process plant is not easy; nor is it something which can be carried out wholly by any one section or department in the purchaser's organization. It must be treated as a joint technical and commercial exercise, and on the technical side must embrace all the technical functions involved in the work concerned. Nor is it simply a matter of assessing capital costs; operating and maintenance charges must also be considered. Further, the effect of financial factors such as terms of payment, the financial consequences of earlier or late completion, and the effect on the purchaser's cash flow position of paying increases in capital costs to secure reductions in operating and maintenance cost, may need to be assessed by the accountants.

The purchaser's overall objective should be to select that offer which he considers will prove to be the most economic when assessed over a reasonable pay-off period, provided always that the capital costs of this offer are such that they can currently be afforded. This assumes that the purchaser, if he is subject to the Public Procurement or Utilities Directive, has stated in his notice in the journal that he intends to award the contract to the firm submitting the most economically advantageous offer. It is not considered that with tenders for plant and equipment or process plant there is ever any justification for the selection of the successful tenderer to be made on the basis only of the lowest price.

It is suggested that in making that assessment it is worth while to systematize the approach, both to establish uniformity and to reduce any bias which there may be towards or against any particular tenderer. The aim should be to make the appraisal as objective as possible. This is a necessary requirement for good contracting practice and mandatory under the Public Procurement and Utilities Directives. It is also necessary under the Directives for an audit trail to be established so that, if challenged, the purchaser can demonstrate objectivity and compliance with the chosen award procedure in his selection process.

## Organization of tender appraisal

In order for the above objective to be achieved the following guidelines are proposed:

1   A formalized procedure should be established and included in the organization's manual of procedures. It is beyond the scope of this work to detail such a procedure but it would need to include:
   - the receipt and administration of the tender documents
   - the responsibilities of the departments involved
   - the setting of objective award criteria
   - the formation of teams for tender appraisal
   - the format of reporting on the appraisal of tenders
   - the establishment, functions and authority of a tender

review board
- authority for the award of contracts
- authorized signatories for contracts.

2    For each contract a team should be established to carry out the appraisal. For tenders of any magnitude it is suggested that this should comprise much the same team who originally carried out the planning of the project (see p. 5) and prepared the enquiry. The team will therefore comprise:

- the project manager as leader
- the project engineer responsible for the technical aspects of the tender – with a multidisciplinary project his task will be to coordinate the specialist engineers each of whom will examine that part of the tender relating to their speciality.
- a purchasing or contract officer who will undertake the assessment of the contractual aspects of the tenders, and
- a representative of the finance department to examine the financial details such as the terms of payment and the effect of escalation formulae. With lower value contracts or where the purchasing/contracts department has the necessary expertise this could be made the function of that department.

Prior to the return of tenders the project manager should have established the plan for the tender appraisal which will comprise:

1    The detailed programme for the appraisal of the tenders, the negotiation with one or more tenderers and the placing of the contract.
2    The availability of the team members.
3    The establishment in detail of the award criteria. If the contract is subject to either the Public Procurement or the Utilities Directives these will have been given in outline in the notice in the journal.
4    Seeing that all administrative arrangements have been put in

place for handling the receipt of tenders, ensuring their secure custody and limited distribution, and accommodation for their secure appraisal in accordance with the appropriate manual of procedure.

As recommended in the list of points for inclusion within the organization's manual of procedure, it is suggested that a tender review board should be constituted which would receive the formal report from the project manager on the results of the appraisal and either give the authority for the award of the contract or for the carrying out of final negotiations.

## Award criteria

If the purchaser has issued a detailed specification of his requirements, it is a reasonably simple matter of checking that the requirements have been met and selecting the firm whose compliant bid is the lowest. If, as is more likely today, the purchaser has issued a functional specification leaving the design responsibility with the contractor, there are two matters which need to be determined. First, for each of the issues which have been identified as being necessary in making the decision on the award of the contract, other than price, it is necessary to establish a minimum below which the bid becomes unacceptable. For example, if the plant is required as a minimum to have an 80 per cent guaranteed availability, then a bid which offered 75 per cent guaranteed availability should be rejected *regardless of price*. Note that the minimum here is the level at which the purchaser is entitled to reject the plant, not that at which he becomes entitled only to the recovery of liquidated damages. The same principle should be applied to the time for completion and to any other criterion which is so significant that a failure to reach the minimum would entitle the purchaser to reject.

Secondly, the purchaser must establish the means by which he is going to relate the non-price elements of the bid, e.g. performance guarantees, reliability guarantees, completion periods, maintenance and servicing obligations, efficiency of equipment (say, energy consumption per tonne of product), differing terms of payment etc., to the tender price in order to obtain an overall comparison between the bids. It can reasonably be assumed that in many instances the bid offering the lowest price will be the bid which offers the optimum in terms of the non-price factors.

The simplest way to do this is to establish in advance of the receipt of the tenders a list for each likely important variable of the values which will be used in making an adjustment to the tenderer's price. These values should in every case be the real cost or benefit to the purchaser and not the values used in any bonus/penalty provisions which are normally for commercial reasons established below the true value of the cost/benefit.

# Methodology

A possible system could be on the following lines:

1  Check the arithmetical accuracy of all tenders. With a plant contract on a lump sum basis the effect of any arithmetical errors will be that the total lump sum does not equal the total of the sectionalized or itemized prices. The often-stated strict rule is that the tenderer should be given the opportunity either to:
   (a)  withdraw his tender, or
   (b)  confirm his total lump sum and indicate the adjustment which he wishes to make to the sectionalized/itemized prices to maintain the arithmetical balance.
   It is for the purchaser, probably through the project manager, to decide whether in any given instance to apply this rule or where there is clear evidence, say that the section price is correct and the lump sum total wrong, to allow the tenderer to correct the total. Where this would be to the purchaser's advantage in that it would be unreasonable to expect the tenderer to stand by the lump sum total, the corrected price would still be the lowest and the purchaser is satisfied that

the error was genuine, then it would seem commercially sensible to allow the tenderer to amend.

2   Consider the total lump sum prices as submitted and establish that each firm has quoted for the same scope of supply. Eliminate from further consideration any offer of, say, more than 20 per cent above the average of the lowest two bids.

3   Have the relevant parts of each tender examined by the team members in accordance with predetermined checklists. An example of such lists for each of the three functions – technical, contractual and financial – is given below. Eliminate any tenders which do not comply with requirements which have previously been established as essential. Adjust each bid by a financial penalty or bonus according to whether or not it would involve the purchaser in additional costs or provide him with extra benefits below or above the previously established norms. In instances where a quantified assessment cannot be made then a qualitative comment should be made.

A simplified example is given in Table 9.1.

4   Reports on each tender to be provided by each function to the project manager for his overall assessment and preparation of his report to the tender review board with his recommendation of the two most favoured tenders in order of priority.

**Table 9.1 Adjusted assessment of tenders for design, supply, installation, commissioning and testing of plant/equipment or process plant**

*Simplified example*

TECHNICAL APPRAISAL                                                  TENDER PRICE
                                                                     ADD DEDUCT

1.1   Tender complies with essential mandatory requirements of the specification?
      If no, the tender should be rejected.                         *yes/no*

1.2   Tender is below required standards in non-essential mandatory requirements of the specification and assessed amount to bring it up to required standards is £......

1.3   Tender is above required standards in the following respects and assessed value of reductions which could be made is £......

1.4    Penalties to be applied due to failure of tender to offer
performance guarantees in accordance with the specification but
which are still acceptable, or bonuses to be applied because
tender offers performance guarantees above those specified.

DESIGN, SUPPLY, ERECTION, TEST AND COMMISSION        TENDER PRICE
AND TEST                                             ADD DEDUCT
1.5    Effect on contract price of alternatives offered, adjusted as
necessary for alterations to programme.
(a)  [*Insert here the items which would be*
(b)  *affected – for example, foundations, struc-*
(c)  *tural steelwork.*]
1.6    Effect of the design offered on the cost of the work to be carried
out by the employer.
1.7    Effect on the purchaser's costs above/below those anticipated
due to:
tenderer's proposed site utilization
tenderer's proposed programme of site works
tenderer's requirements for the use of common facilities with
other contractors.

DESIGN, SUPPLY, ERECTION, TEST AND COMMISSION        TENDER PRICE
AND TEST                                             ADD DEDUCT
1.8    Assessment of costs which will be incurred by the purchaser due
to:
items excluded by the tenderer from his scope of work
demands made by the tenderer on the purchaser for the
provision of extra facilities for testing etc.
location of contractor's works causing extra costs for contract
administration, visits to inspection etc.
delays in tenderer's response time to the remedying of defects
due to his remote location
tenderer's spares recommendations being above the anticipated
level
consumption of consumables being above the anticipated level.
1.9    Capitalized effect of additions to, or deductions from, the stated
norm for operating labour. Effect to be assessed over, say, ten years.
1.10   Capitalized effect of any additions to, or deductions from, the
norm of maintenance costs due to equipment or other work
standard offered by the tenderer as part of his specification – for
example, use of pumps with low initial but high operating costs,
painting of steelwork to reduced standards. The effect to be
assessed over, say, ten years.

DESIGN, SUPPLY, ERECTION, TEST AND COMMISSION        TENDER PRICE
AND TEST                                             ADD DEDUCT

1.11   Does the tender meet the minimum performance standards
specified by the employer in his inquiry?              *yes/no*

1.12 If yes, does the tender guarantee any financial benefit to the
employer over the minimum standard specified?        yes/no
If yes, state the assessed benefit capitalized over, say, ten years,
taking into account any additional expense to which the employer
would be put to earn such benefit.

1.13 Has the tenderer accepted the liquidated damages specified
for failure to meet guaranteed performance?        yes/no

1.14 If no, state the capitalized detriment the employer would suffer by
acceptance of the tenderer's proposals for a given loss in efficiency.

COMMERCIAL                                       TENDER PRICE
                                                 ADD DEDUCT

2.0 Has the tenderer made any qualifications to the proposed
contract conditions? If so, assess the additional risk/cost to the
purchaser if these were accepted. Examples could be:
      inclusion of overall limit of liability
      reduced defects liability period
      exclusion of liability for defects after
      expiry of defects liability period
Addition of extra events allowing the tenderer an extension of
time for completion
Reduction in rate of liquidated damages or lower limit of liability
Exclusion of liability for delay after maximum limit of liquidated
damages reached
Reduction in liquidated damages for failure of plant to meet
performance requirements
Limitations on purchaser's right to reject if plant performance is
below a level at which maximum damages are reached.

2.1 Has the tenderer agreed to satisfy the requirements in the
invitation to tender regarding the submission of bonds and
parent company guarantees? If not, are any modifications
proposed acceptable?        yes/no

2.2 If the tenderer is a consortium or joint venture, is it clear that
all members accept joint and several liability?        yes/no
Note. If the answer to either 2.1 or 2.2 is no, the tenderer must be
required to amend.

2.3 If the tenderer is an overseas firm, has he quoted on a totally
inclusive basis for all costs involved in delivering material to site
and bringing in of any foreign labour or supervision? If not, any
extra costs must be assessed and added.

FINANCIAL                                        TENDER PRICE
                                                 ADD DEDUCT

3.0 Has the tenderer quoted in the required currency, normally
sterling, without reference to an exchange rate?        yes/no
Note. If no, it is suggested that the tenderer should be required
to agree to his tender being converted at the exchange rate ruling
at the date of tender submission and thereafter to remain fixed,
or to withdraw his tender, unless the purchaser is willing to
accept the exchange risk. In the latter event the purchaser must

make an assessment of his additional risk and add it to the
tender price.

3.1   Has the tenderer quoted on a fixed price basis or, if the
      enquiry allowed for escalation, in accordance with the
      formula proposed by the purchaser?                    *yes/no*
      If no, it is suggested again that the tenderer should be required
      to conform to the terms of the enquiry or withdraw unless the
      purchaser is willing to accept the additional costs, in which event
      he must make an assessment and add it to the tender price.

3.2   Has the tenderer accepted the proposed terms of
      payment?                                              *yes/no*
      If no, again it is suggested that the purchaser should proceed as
      in 3.1 above.
      If tenderers have been asked to put forward their own proposals
      on terms of payment, the purchaser must bring these to a
      common basis for appraisal purposes by selecting the one which
      is the most favourable to him and adjusting the others.

The above table should be completed for each tenderer in a
standarized format. Although the heading is 'Technical
appraisal' the adjustments to be made to the price should be the
joint decision of the engineering and commercial staff engaged
in the appraisal.

The above notes suggested that the adjustments where
necessary to the tenderer's price should be made by the
purchaser. It is recognized that some organizations proceed in
the alternative manner of asking the tenderer to price out the
qualification which he has made himself. The risk in proceeding
in that manner is that it encourages the tenderer to put in the
qualification so as to give himself the chance of either adjusting
his price or not after the bids have been opened, and when he
can be assumed to have a reasonably good idea of where he
stands in the order of bids. For this reason it is considered that
the purchaser should make the adjustments himself in an
objective manner which is capable, if necessary, of being justified
as fair to the tenderer.

The system of tender appraisal referred to above should be
sufficient to cover most cases in which the contract period does
not exceed say 18 months to two years. For contracts exceeding
that period, and where the offers received differ significantly in
respect of the terms of payment and CPA formulae proposed,

then the employer should consider the additional step of making a comparison on a common financial basis.

The way to do this is to convert each of the anticipated payments which the employer will have to make into its Nett Present Value (NPV) and by adding these arrive at the NPV of the tender. The NPV of each payment is calculated by discounting the cash value of the payment back to its present value through the use of an assessed discount factor. That factor is a combination of an assumed interest rate, often taken as the mean cost of capital to the employer, applied over the period of time from when the payment has to be made, back to the date at which the calculation is being performed. The formulae for calculating the NPV of a sum of money $S$ in any year $n$ at a discount rate of $r$ is

$$\frac{S_n}{(1 + r)_n}$$

The actual arithmetic can be done either using discount tables or by a modern calculator.

Applying the concept of discounting, payments which have to be made early will cost the employer more than those which have to be made later. To illustrate the effect of discounting if, under the terms of the tender, a payment of £300 000 was due at the end of three years from date of contract, with a discount rate of 10 per cent this payment would have an NPV of £225 394.

It was suggested in the previous edition that, having adjusted the prices of the tenderers, a further exercise using points could be adopted so that the various factors of price, delivery, management, quality and reliability could be assessed to give an answer which took account of their weighting and of factors which it is difficult to quantify. Although points systems have been criticized it is still considered that where non-price factors such as quality of staff and management and track record of achievement are particularly important, as in the selection of turnkey contractors, management contractors and consultants, the use of a points system is advantageous.

If a points system is to be adopted, the following issues need to be decided in advance:

1 The factors other than price to be taken into account.
2 The respective weighting to be given to price as compared with the other factors.
3 The respective weightings to be given to each of the non-price factors as between themselves.
4 How the points are to be scaled within each factor.

Finally, assuming that the price has been adjusted to take account of technical non-compliance and other factors as indicated in Table 9.1, care must be taken to ensure that there is not double counting by taking those factors into account again when assessing, say, technical merit or completion.

In essence the points system is a method of relating the quantitative element of price with the qualitative elements of management skills, reliability and the probability that the firm will complete the contract to time, budget and performance.

An example is given below:

Assume that three tenders are being finally considered. The values of each, adjusted to take into account all the factors referred to in Table 9.1, are *A* £184 000, *B* £187 000, *C* £191 000.

All firms have accepted the completion period of forty weeks, but based on discussion held with the tenderers and on past performance, it is considered likely that *A* will take forty-four weeks and *C* forty-two. Again based on discussions and past performance, it is considered likely that *B* will in fact complete on time.

The proposals submitted by the firms on management and control are all reasonable, but those by *B* are likely to ensure better co-ordination and less risk of other delays or claims.

Based on recent experience, the Q & R assessment of the three firms is that on the last contracts they completed of a similar nature: *A*'s plant suffered minor troubles causing a week's delay in going into operation; *B*'s plant had one large breakdown due to a failure of a sub-contracted item; *C* had a number of mechanical troubles which took over three months to correct.

Time is considered important, and it is decided in advance of going out to tender to adopt the following points system for comparing the offers.

## Price

Thirty points to be awarded to the lowest tenderer, the others to have 2.5 points deducted per £1000 in excess of the lowest offer.

## Completion

Twenty-five points to be awarded to the firm which it is decided, on the basis of their offer and in discussion held with them, will achieve the best delivery date; five points to be deducted per week in respect of the difference between that period and the completion periods which it is decided are likely to be met by the other tenderers.

## Management

Twenty points to be awarded, based on an assessment of the proposals submitted.

## Q & R

Twenty-five points to be awarded, five points to be deducted for each week's delay which it is considered likely that the plant submitted by any tenderer will suffer in going into operation.

On the basis of the above, the comparison between the three firms would look as follows:

| Item | A | B | C |
|---|---|---|---|
| Price | 30 | 22.5 | 12.5 |
| Completion | 10 | 25 | 15 |
| Management | 15 | 20 | 15 |
| Q & R | 20 | 20 | 10 |
| | 75 | 87.5 | 52.5 |

The result of the above exercises would be quite clearly a decision that the contract be awarded to firm *B*. It is essential, if objectivity is to be maintained when using a points system, that the weighting to be given to the different factors of price, completion etc. should be decided upon in advance of the submission of the offers.

In the case of building and civil engineering contracts the procedure will differ in that the work of tender appraisal will normally be largely the responsibility of either the employer's own civil engineers or quantity surveyors or consultants employed on his behalf. However the following points do require attention on the commercial side:

1   A civil engineering contract under the standard ICE conditions is a re-measurement contract; there is no initial lump sum price. If therefore there is an error in extension this has no effect on the final price paid by the employer. It is accordingly necessary for the individual rates to be checked. This is also a safeguard against the submission by a tenderer of an unbalanced bid in which he has priced some work high, and other work low, in the belief that there will be a substantial increase in the quantity of some and a decrease in the quantity of others. Any such bid should be rejected.

2   On lump sum contracts if an error in rates, extensions or totals is not discovered by the employer or the engineer before a contract is awarded the contractor is bound to carry out the original work at the tendered sum. If however the employer or engineer does discover the error through reading the bills of quantity (which in this case are only to be used for the purpose of pricing variations), the courts would order rectification of the error, so the tenderer ought to be allowed the opportunity to correct the mistake. This means that on a lump sum building contract at least the bills of quantity of the lowest two tenderers ought to be checked, particularly if these are close together in price, in order to ensure which of them is the lowest.

    This assumes that the error is genuine and not a deliberate mistake by the firm so it can have the opportunity either to correct it or not, once it knows the prices of the other bidders. If that is suspected, the firm should be told either to stand by their tender or withdraw – see the *Code of Procedure for Selective Tendering* published by the National Joint Consultative Committee of Architects, Quantity Surveyors and Builders.

It is important that the contracts or purchasing officer as representing the employer should be aware of, and involved in, these issues, since the engineer has normally no authority on the employer's behalf to make decisions relative to mistakes at the tendering stage. For a clear and detailed discussion of these issues and the ICE form of contract generally see Max W. Abrahamson, *Engineering Law and the ICE Contracts* (Applied Science Publishers, 4th edition, 1979).

3   By virtue of the risks involved in the design and execution of civil engineering works, and of the way in which through the conditions of contract these are apportioned as between the employer and the contractor, there is a strong tendency for such contracts to become a battlefield for claims rather than a cooperative effort between the parties to achieve their common objective of completing the works to the employer's satisfaction and of the contractor being fairly rewarded for his efforts. Much can be done at the stage of inviting tenders and of tender analysis to improve this situation, first by the careful selection of firms to be invited to tender and then by:

- ensuring that the tenderers have made available to them all information necessary relating to the physical conditions likely to be encountered and the requirements of the employer and the engineer relative to the design and execution of the works; and
- examining the initial low bidder sufficiently in respect of his construction methods, sources of materials and labour, plant availability, construction programme, intended site management and his design proposals for temporary works, as to minimize his opportunities for the submission of claims and satisfy the employer and his engineer that the contract is likely to be properly and efficiently implemented.

Blind acceptance of the apparently lowest offer is only likely to result in an over-run of the cost budget, delays in completion and endless hours spent in wrangling.

When the final steps of the analysis procedure have been taken there are three possibilities:

(a) there is one bid which is in conformity with the purchaser's requirements and which he is prepared to accept without further negotiation, or
(b) there is one bid which the purchaser prefers significantly to any other but which does not wholly meet his preferred requirements, or
(c) there are two or more bids which are close enough to each other that the purchaser would prefer not to make a decision until after further negotiation.

Where the purchaser is subject to the Public Procurement Directives, even if he has chosen the restricted procedure, it would appear that it is not open to him to undertake post-tender negotiation. It has been stated by the Council and the Commission that:

in open and restricted procedures all negotiations with tenderers on fundamental aspects of contracts, variations of which are likely to distort competition and in particular on prices shall be ruled out; however discussions with tenderers may be held only for clarifying or supplementing the content of their tenders or the requirements of the contracting authority provided this does not involve discrimination.

With contracts subject to the Public Procurement Directives therefore placed under the restricted procedure, it would appear that the purchaser must accept the most economic offer provided that it meets his mandatory requirements as specified in the invitation to tender, even if he believes that he could obtain a better bargain by post-tender negotiation. In particular it is suggested that he cannot seek by such negotiation to obtain a reduction in the tender price. Although there are provisions in those directives under which, exceptionally, tenders may be invited according to the negotiated procedure, they are only of very limited application – for details see Trepte, *Public Procurement in the EC*, CCH, 1993, p. 121 et seq. For present purposes they will be ignored.

However the Utilities Directive does allow the purchaser an absolute freedom to choose the negotiated procedure. In this instance therefore it does not appear that there are any restrictions on the purchaser's right to negotiate, provided that he does not offend against the basic rules of objectivity and equality of treatment. With contracts not covered by any of the directives the purchaser is totally unrestricted in his entitlement to negotiate.

Therefore under the Utilities Directive or with contracts not covered by any directive it is suggested that the purchaser in cases (b) and (c) above should proceed to negotiate. Only in the very limited circumstances that the purchaser is regularly in the market for the work in question, the number of firms with whom he deals for that work is limited and they are all confident that the purchaser *never* engages in post-tender negotiations, will the tenderers have followed the rule of 'final offer first'. In any other case the tenderers in order to protect themselves will have included items of 'fat' in their bids in order to have something to give away, if necessary, in negotiations. The implications of this to the purchaser are clear. Unless he can be totally confident that he is in the one case above described, when he will have received the firm's best offers the first time round, then he should negotiate – and he should do so in case (c) above with both tenderers who should each be aware of the negotiations taking place with the other.

Obviously a careful record must be kept by the purchaser of any negotiations held and of price reductions or other amendments agreed by the tenderer to his tender. After the conclusion of the negotiations, unless prior approval has been obtained, the project manager as leader of the negotiations should refer back to the tender board for authority to award the contract. There are then still two important steps to be taken.

First, a permanent record must be made of the contractor's success or otherwise in bidding that particular contract. This should record the salient features brought out by the tender appraisal, that is:

1   Price at which the contract is placed, or would have been placed if the tender had been accepted.
2   Completion period promised related to that price.
3   If applicable, the performance guaranteed.

Ideally these items should be recorded in such a way that they can at a later date be compared in the case of the successful tenderer with the same data derived from the contract completion report. In this way an assessment can be built up of what was achieved against what was promised at tender stage. This information, together with the data on those firms who were unsuccessful, can then in turn be used to build up the vendor rating assessment for use in selection of firms to go on future tender lists and in the appraisal of offers when submitted.

The whole operation becomes a continuous cycle. It is of course necessary also to try to avoid making it a closed shop of a slowly diminishing number of firms. Assuming that the level of demand for the particular types of work involved remains at least partly static, the employer must ensure that he is continually testing the levels of price, delivery and quality by inviting new firms that he considers capable of meeting his standards.

Finally, if the firms that are unsuccessful are to be given the chance to improve their performance, they must be told where they went wrong. Once, therefore, the contract has been placed, each of the firms that were included in the final shortlist should be given the opportunity to come and discuss their bid, and the points where it was considered to be unfavourable should be brought out in these discussions. It must be made clear that the purpose of the meeting is to permit the firm to improve its performance on the next occasion, and there must be no question of jobbing backwards, nor should the discussion be allowed to become the occasion for a criticism of the buyer's decision.

Having taken the trouble in the example quoted above to discuss with *A* and *C* where their tenders were marked down and why, the buyer must then do two more things:

1   Before asking them to tender next time, insist that they satisfy him they have taken steps to remedy their weaknesses.
2   When so satisfied give them a chance to tender again.

On the tender analysis the next time the tenders would of course be judged on their merits as then presented, plus the buyer's assessment of the firm's current performance level. He should not take into account the old faults, which by putting them on the current tender list he is accepting have been put right. 'Give a dog a bad name' is too common a failing in the contracting industry, and firms continue to be penalized for errors made years ago under different conditions, and often under different management, which should long since have been treated as wiped out.

# 10 Placing the contract

Previous chapters have dealt with the planning of the contract, the invitation and submission of tenders and the appraisal of competing offers. Once the selection of the successful tender has been made and authority given by management to go ahead with the contract, there will be strong pressure for instructions to be given to the contractor for work to be started immediately and in advance of any formal contract documentation. The contracts officer faced with such pressure is often in a difficult position. On the one hand he knows that to delay starting work for the sake of 'getting the paper straight' can cause a genuine delay to the project and increased expenditure. On the other hand he is also aware of the dangers of allowing the contractor to proceed without having the loose ends tied up, and the weakness of his own negotiating position relative to the contractor once the latter has been authorized to start work.

By taking preventive action in advance there is much the contracts officer can do to avoid or minimize the risk of getting caught in this situation. Some suggestions are as follows:

1   Wherever possible, issue the inquiry in such a form that the

tenders when submitted are likely to be complete and constitute an offer capable of being accepted with the minimum of amendment.

2  If the tender is not wholly acceptable, negotiate with the tenderer in advance of receiving the go-ahead from management. Looking at Table 1.2 (p. 13), the buyer could well use week 14 while the works manager is obtaining financial approval to complete his negotiations and have the order all agreed with the supplier and ready to issue by the time the authority is received.

3  Do not invest the placing of the order or contract with undue solemnity. It should not for instance require more than one signature.

4  Do not try to obtain the ultimate in the completeness or comprehensiveness of the contract document at the expense of never finalizing the draft. To wait until there are no changes pending to the specification may mean waiting until after the plant has been built.

It may be suggested to the contracts officer that his problem could be solved by issuing the contractor with a letter of intent. The difficulty with letters of intent is to ensure that both parties know and understand precisely what they mean. It is fundamental to English contract law that there can be no lesser legal obligation than one which is contractually binding. Either, therefore, the letter of intent constitutes a contractual commitment, for the breach of which an action for damages would lie, or it is merely an expression of intention which is legally unenforceable – remember the discussion on p. 72. In the preparation of a letter of intent, therefore, one needs to be absolutely clear what is meant. It may well be, for instance, that the intention is to give an indication to the contractor of one's intention to proceed with the whole job, but with no contractual commitment to do so, whilst at the same time authorizing him to incur certain specific preliminary expenses which would constitute a definite commitment.

Such a letter is more correctly called an instruction to proceed (ITP) but in general commercial practice is still often referred to as a letter of intent.

An example of an ITP or letter of intent which is intended to have a limited contractual effect might read as follows:

I am writing to confirm that it is our company's intention, subject to the satisfactory conclusion of negotiations between us, to place a contract with your company for the design, supply, construction, and commissioning of . . . . . . . . . . . . . . . . . . . . . . . . . . . . . . . . . . . . . . . . . . . . [for the sum of £ . . . . . . . . . . . . (insert if already agreed)]. The contract will be generally in accordance with the terms set out in your tender dated . . . . . . . . . . . . . . . . . . . . . other than for the clauses set out in Annexe I hereto which still remain to be agreed between us. The programme for finalizing all outstanding issues between us is set out in Annexe II.

Pending the conclusion of our negotiations you are hereby authorized to proceed with preliminary design work for the contract in accordance with such instructions as you may receive from . . . . . . . . . . . ., our Chief Engineer, up to a total value not exceeding £ . . . . . . . . . . . . priced at the hourly rates for design staff set out in your Tender.

You are also authorized to purchase the long lead items listed in Annexe III hereto at the prices stated therein.

On the placing of the contract with you all work carried out by you under this letter of intent will be deemed to have been carried out by you under the terms and conditions of the contract.

If we are unable to reach agreement with you on the outstanding issues between us within a period of . . . . . . . . weeks from the date hereof we shall have the right to terminate this letter of intent by notice in writing. In that event:

(a)  we would reimburse you for the design work carried out by you under the terms of this letter up to the date of termination to a limit of £ . . . . . . . . . . . . together with the cancellation costs reasonably incurred by you in respect of the orders for the long lead items. Alternatively we would have the right to take over such orders from you, and all orders placed by you shall include such provisions.

(b)  the property in all drawings and other documentation prepared by you under the terms of this letter and any materials manufactured would vest in us.

Please acknowledge your acceptance of this letter and confirm that you will be starting work immediately.

The actual form of the contract documents as such will depend largely on how the tendering has been carried out and whether the tender as received is suitable for acceptance.

If a formal inquiry was issued, complete with terms and conditions, then:

1   If the tender as received is suitable for acceptance with no qualifications, the contract can be placed by a simple letter of acceptance.
2   If the tender as received cannot be accepted without amendment, either:
    (a)   if there are only a few amendments, they can be set out in the letter of acceptance and the tenderer can be asked to confirm his acceptance of these, or
    (b)   if the amendments are more extensive, the tenderer can be asked to resubmit his tender so that the procedure in (a) above can be followed.

If the inquiry was not complete or if very substantial changes are required as a result of post-tender negotiations, it will usually be more convenient for these to be incorporated into a single contract document.

One trap to be avoided is that of attempting to incorporate within the contract post-tender agreements reached between the parties, by either annexing to the letter of acceptance copies of correspondence or minutes of meetings or identifying them in a schedule. Invariably such correspondence and/or minutes will be partially contradictory and contain matters which were never considered at the time by the parties as contractual obligations. The task then of interpreting objectively from a study of such documents just what it is that the parties must have intended to be their respective obligations is often a matter of great difficulty. At the very worst it could lead a court to conclude that since they cannot decide just what the bargain was that the parties believed they had made, in fact they never made one at all, and there is no contract.

There is no particular merit or legal significance in the form which the contract takes, unless it is desired by the purchaser to have the contract executed as a deed and so obtain the benefit of the 12-year prescription period for breach of contract rather than the 6-year period which applies to contracts executed under hand. This is really the only benefit which is gained by the use of a formal agreement and the only justification for having one prepared, unless of course it is required by the standing orders of the authority. In any other instance there seems absolutely no advantage to be gained in accepting a tender by letter and then having a formal document prepared. This is really a complete waste of time and effort. The aim should be at all times to keep the contract documentation as short and simple as possible consistent with clarity of meaning.

An example of a simple letter of acceptance would be:

I am pleased to inform you that the . . . . . . . . . . . . . . . . . . . . . Company Limited hereby accept your tender dated . . . . . . . . . . . . . . . . . . . . . . . . . . . . for the design, supply, construction, and commissioning of a . . . . . . . . plant for the fixed lump sum of £. . . . . . . . . . . . . . .

The engineer appointed for this contract is . . . . . . . . . . . . . . . . . . . . ., the Company's Chief Mechanical Engineer. You should contact him immediately for instructions to start work. You should forward immediately to the Engineer the following documents all as specified in the contract conditions:

the insurance policies
the parent company guarantee
the performance bond.

Please acknowledge receipt.

If desired, any particular instructions in invoicing could be added as an additional paragraph, but are probably best dealt with in a separate letter or in general notes on administration.

Note that the contractor is only being asked to acknowledge receipt, not to 'accept' the letter, since, assuming that the letter is issued during the tender validity period, the contract is created as soon as the letter is posted.

If there are one or two modifications or amendments to the tender, the letter might read:

I am pleased to inform you that the . . . . . . . . . . . . . . . . . . . . . Company Limited hereby accept your tender dated . . . . . . . . . . . . . . . . . . . . . . . . . for the design, supply, construction, and commissioning of a . . . . . . . . plant, subject to the following:

1   Inclusion of Alternative *A* on page 5 of your Tender. This means that the contract price will now be the fixed lump sum of £. . . . . . .
2   Deletion of the price escalation clause. The contract price is fixed against any changes in costs.
3   Reduction of the period for completion from thirty-six to thirty-two weeks.

The engineer appointed for this contract is . . . . . . . . . . . . . . . . . . . . . ., the Company's Chief Mechanical Engineer. You should contact him immediately for instructions to start work. You should forward immediately to the Engineer the following documents all as specified in the contract conditions:

the insurance policies
the parent company guarantee
the performance bond.

Please confirm your acceptance of the above.

Note that in this case the contractor is asked to confirm his *acceptance* since his offer is not being 'accepted' entirely in the terms in which it was made. The contract will only be formed, therefore, when the contractor sends his unconditional acceptance of the above. It may be convenient to issue this letter in duplicate with a space for the contractor to sign and return the duplicate as agreed, provided the amendments have already been informally agreed with him. This avoids the possibility that he may when replying use a standard form which refers to conditions different from those which apply to this contract.

If he were to do this, it would be a question of having to decide whether the accompanying letter amounted to a counter-offer or not. Just two of the main cases on what is often referred to as 'the battle of the forms' will be mentioned in order to illustrate the perils involved. In the first (*Butler Machine Tool Co.*

*Ltd v Ex-Cell-O-Corporation (England) Ltd* [1979] 1 All ER 965) the seller returned to the buyer the tear-off printed acknowledgement slip which was part of the order and read 'We accept your order on the terms and conditions stated thereon'. However, he did so with a covering letter which stated that the order was 'being entered in accordance with our revised quotation of 23 May'. Not surprisingly the terms of that quotation differed substantially from those of the buyer. The Court of Appeal held that the buyer's order was a counter-offer which the seller accepted by returning the acknowledgement slip. The accompanying letter was held to be irrelevant; it merely referred to the identity and delivery period for the goods.

By contrast in *Muirhead v Industrial Tank Specialities* [1986] 3 All ER 705, the seller used *his own* acknowledgement slip which itself stated that 'We thank you for your order which will be executed in accordance with our general conditions of sale (see over)'. The court held that the acknowledgement slip constituted a counter-offer which was accepted in due course by delivery being taken of the goods. Accordingly the contract was on the seller's terms which from reports of the pre-contract discussions between the parties and indeed the way in which the order had been prepared was probably not at all what, subjectively at the time, had been intended. However as must be stressed, because the point is often missed, the position under English law is that 'an offer falls to be interpreted not subjectively by reference to what has actually passed through the mind of the offeree, but objectively by reference to the interpretation which a reasonable man in the shoes of the offeree would place on the offer' per the Court of Appeal in *Centro-Provincial Estate v Merchants Investors Assurance Company* [1983]. Evidence of the party's subjective intentions in the matter of formation and indeed of contract interpretation generally is therefore irrelevant.

If for the reasons indicated above it is necessary to have a formal contract, this should still be as short and simple as possible. Ideally the contract document should consist of about seven clauses defining the basic obligations of the parties with everything else contained in schedules. A suitable layout would be as in Table 10.1.

**Table 10.1    Form of agreement**

CLAUSE
1   Description of contract work.
2   Work to be done and services to be provided by the purchaser.
3   Contract price.
4   Programme/time for completion.
5   Performance guarantees.
6   Appointment of engineer.

SCHEDULE
A   Specification.
B   Contract drawings.
C   Schedule of prices.
D   Programme.
E   Guarantees.
F   General conditions of contract.

# Contract work

A suitable draft paragraph for a substantial plant contract might be:

The work which is the subject of the Contract comprises the design, supply, erection, testing, and commissioning on one . . . . . . . . . . . . . . ., with all necessary ancillary equipment as described in the Specification (schedule *A*) and in the Drawings (schedule *B*).

One problem which may arise in defining the contract work is where the purchaser has issued a specification with the inquiry which defines the performance required of the plant and the standards to which it is to be designed and built, and the contractor in tendering has put forward a detailed specification of what he is offering to meet these requirements. There are, therefore, two specifications. It is important first to check that there are no discrepancies between the two specifications, for example different terminal points, reference by the bidder to his assuming the purchaser will supply storage *accommodation*, whereas the purchaser has only stated he will allocate storage *space*, etc.

Frequently there will be technical discussions between the purchaser's and contractor's engineers to remove minor discrepancies and incorporate any late changes in thinking, or possibly make savings to bring the contract price below budget. These changes to the specification will usually have been recorded in letters or notes of meetings. As referred to earlier the only safe way of incorporating them into the contract is to make the amendments to the specification itself. Indeed just doing this will frequently reveal other necessary consequential changes and also show up any ambiguities in the drafting.

Second, it is important that in accepting the contractor's tender the purchaser should make it clear that he is not taking any responsibility that what the contractor is offering will in fact meet the purchaser's requirements. For example, the purchaser may have specified a conveyor capable of performing a certain duty. The tenderer may include in his offer a description or drawing of a certain design feature the inclusion of which in fact makes it impossible to achieve that duty, although this may not be discovered until completion. When this happens, and the defect is discovered on testing the conveyor on site, the purchaser wants to be in a position to reject the plant until the defect is remedied. To be certain on this point, it would be advisable for the purchaser to word his contract along the following lines:

design, supply, erection, and commissioning of the plant as described in the contractor's specification (schedule *A* part I) provided always [and this is an essential condition of the contract for which the contractor is responsible and on which the purchaser is relying wholly on the contractor's skill and judgement] that such plant satisfies in all respects the terms of the purchaser's specification (schedule *A* part II).

This would then clearly bring the contractor's obligations within the scope of the words from Hudson's *Building and Engineering Contracts* (1959), 8th edn, p. 147, summarizing a long line of English cases which were quoted with approval in the decision of the Supreme Court of Canada in *Steel Company of Canada Ltd v Willand Management Ltd [1966]*

Sometimes again a contractor expressly undertakes to carry out work which will perform a certain duty or function in conformity with plans and specifications and it turns out that work constructed in accordance with the plans and specifications will not perform that duty or function. It would appear that generally the express obligation to construct a work capable of carrying out the duty in question overrides the obligation to comply with the plans and specifications and the contractor will be liable for the failure of the work notwithstanding that it is carried out in accordance with the plans and specifications.

## Purchaser's obligations

A suitable clause defining the purchaser's obligations might be:

The purchaser is responsible for carrying out the work and providing the services set out in Appendix . . . . . . . . to the specification and for ensuring that these are carried out/provided at the times stated in the programme or, where no times are so stated, at such times as will enable the contractor to comply with his obligations under the contract.

Even in the absence of these words there is an implied obligation on the part of the purchaser that the services have to be provided by him at a reasonable time (see p. 64).

It is convenient to bring together all the purchaser's obligations as regards work and services in one schedule so that this forms a checklist for the contracts officer and engineer administering the contract. It should ensure that arrangements are made well in advance for these items to be provided. The time factor is almost as important as the service itself. It is not much use making ground available for storing steel sections after the steel has been delivered and the contractor has had to find room for it somehow within the working area. The purchaser who does this has only himself to blame when he gets a claim for double handling and loss of productivity.

# Contract price

The definition of the contract price will depend on how the price is to be determined. The methods of doing this are discussed in detail in Chapter 13.

If the contract price is a lump sum, the clause can be very simple, for example:

The purchaser shall pay the contractor the lump sum of £. . . . . . . . . . . ('the contract price') plus or minus such other sums (if any) as under the contract are to be taken into account in ascertaining the contract price.

If the contract price is to be determined according to the value of work done, using a bill of quantities or schedule of rates, the clause might read:

The purchaser shall pay the contractor the value of the contract work executed in accordance with the contract ('the contract price') as determined by the engineer/architect by measurement of the work done and valuation of the same at the rates and prices set out in the contract plus or minus such other sums (if any) as under the contract are to be taken into account in ascertaining the contract price.

If the contract is wholly or partially on a cost re-imbursement basis or target cost, the assessment of the contract price becomes that much more difficult. The important points which have to be covered are set out in detail in Chapter 13, pp. 229–35.

# Programme/time for completion

This must tie up with the rest of the contract so that there is no ambiguity as to what is meant by 'completion'. On a plant contract there are two alternative approaches which can be adopted. The first is that the tests on completion are actually included within the definition of completion as in MF/1 where there are two separate obligations: to complete the works according to the contract *and* to carry out the tests on completion by the time fixed for completion – see clause 29. The alternative is to provide as is often found in process plant conditions that the obligation is 'to complete the works ready for the carrying out of the take-over tests' by the time fixed for completion. Obviously there is a very significant difference between the two and the agreement must set out whichever is intended. An example might be:

The contractor shall complete the construction and testing of the works so as to be entitled to apply to the Engineer for a Taking Over Certificate under clause . . . . . . . . . . . . . . of the General Conditions of Contract not later than . . . . . . . . . . . . . . . . . . . ('the date for completion') or any extension of that date to which the contractor may be entitled under the contract.

## Performance guarantees

This need be no more than a simple statement that the contractor guarantees that the plant will meet the performance guarantees. Suitable wording would be as follows:

The contractor undertakes that the works will meet the guarantees set out in the . . . . . . . . . . . . . . . . schedule when tested in accordance with the test procedure set out in the specification and after allowing for the specified margins of tolerance.

## Conditions of contract

Conditions of contract are often conveniently described as being either 'general' or 'special'. General conditions are those which are set out in standard forms prepared either by one of the engineering institutions, for example ICE Conditions of the Institute of Civil Engineers, or the form MF/1 of the Institutions of Electrical and Mechanical Engineers. Special conditions may be required, either because of some issue not dealt with in the general conditions or because the purchaser wishes to have the general conditions modified in certain respects.

An example of the first would be modifications to the Clause 35 of the MF/1 Conditions if on the particular project the Performance Tests were to be carried out prior to Take Over. An example of the latter would be clauses relating to the provision by the Contractor, if a subsidiary company, of a Parent Company Guarantee, Prevention of Corruption and Approval of Publicity material which as a standard are added to the ICE Conditions of Contract 6th edition by the National Rivers authority.

With general conditions, it is normally only necessary to refer to them in the schedule. All the standard forms now contain a schedule or appendix listing such items which must be completed by the purchaser, otherwise it will become impossible to give effect to the contract conditions to which such items relate. If a purchaser is habitually placing contracts incorporating a certain set of general conditions, it is advisable to have the schedule pre-printed with the references to the clause numbers already included so that nothing is overlooked.

Special conditions must of course be set out in the schedule in full. Care should be taken to see that they are consistent with the general conditions, that is that words are given the same meaning and the same words are used to describe the same item or activity. For example, if the general conditions use the expression 'take over' when referring to the point at which the purchaser assumes responsibility for the plant, then the special conditions should likewise use 'take over' and not 'acceptance'. It is a rule of construction that if a draftsman has used two different words he will be assumed to have done so deliberately, and that therefore they have different meanings.

# Appointment of engineer or architect

It is usual in UK-based contracts, or where a UK consultant is employed, to appoint an engineer or architect on a building contract to represent the purchaser. His functions and powers are described in Chapter 21. This can be done quite simply by stating that:

The engineer/architect appointed by the purchaser for this contract is
.......................... or the person whom the purchaser may
subsequently notify to the Contractor in writing.

The full draft of the contract document might then be as follows:

This agreement is made the .............. day of ...............
19....... between ........................ (the purchaser) of the
one part and ............................. (the contractor) of the
other part. Whereby it is agreed as follows:

## The contract work
The work the subject of the contract comprises the design, supply,
erection, testing, and commissioning of one ................ complete
with all necessary ancillary equipment in accordance with the
Specification set out in schedule *A* and the Drawings referred to in
schedule *B*.

## Work to be done and services to be provided by the purchaser
The purchaser is responsible for carrying out the work and providing
the services set out in Appendix ........... to the Specification and
for ensuring that these are carried out/provided at the times stated in
the programme (schedule *C*) or where no times are so stated at such
times as will enable the contractor to comply with his obligations
under the contract.

## Contract price
The purchaser shall pay the contractor the fixed lump sum price of
£............ ('the contract price') plus or minus such sums (if any)
as under the contract are to be taken into account in ascertaining the
Contract Price.

## Programme/time for completion
The contractor shall complete the construction and testing of the works
so as to be entitled to apply to the Engineer for a Taking Over
Certificate under clause ........ of the general conditions of contract
not later than ................ ('the date for completion') or any

extension of that date to which the contractor may be entitled under the contract.

## Performance guarantees

The contractor undertakes that the works will meet the guarantees set out in the . . . . . . . . . . . . . . . . . . . . . schedule when tested in accordance with the test procedure set out in the specification and after allowing for the specified margins of tolerance.

## Conditions of contract

The contract shall be carried out under the general conditions of contract referred to in Part I of schedule *D*, the special conditions of contract set out in Part II of that schedule and the particulars stated in the Appendices thereto.

## Engineer

The Engineer appointed by the purchaser for this contract is . . . . . . . . . or the person whom the purchaser may subsequently notify to the contractor in writing.

# Part Three
# TERMS AND CONDITIONS OF CONTRACT

# 11 Standard terms of contract: I

## Interrelationship of conditions of contract

Conditions of contract are included within the contract to express the relationship between employer and contractor and to define explicitly what is to happen should that relationship be disturbed by the failure of either party to fulfil their obligations. To this extent they are a reflection of the practicalities of the contract work. When, for example, reference is made in them to 'completion' this is not some abstract legal concept but the very fact of the 'topping-out' ceremony on a building or of the anxieties of the moment when a process plant first goes on stream. The legal requirement should always be a reflection of the practical possibilities. Escape from that and the contract conditions become at best a sterile exercise in drafting and at worst an infliction of penalties upon the innocent and unwary.

No matter what the subject matter, all engineering contracts have the same basic framework, no part of which can be altered or omitted without it affecting at least one other part. The basic framework is illustrated in Figure 11.1, which is in the form of a network analysis.

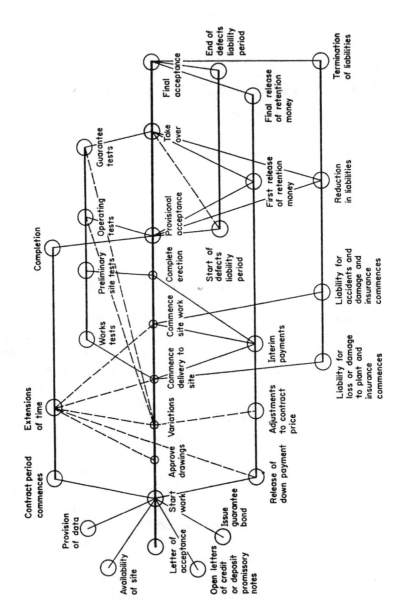

**Figure 11.1   Network analysis**

Solid lines indicate that two events will always be interdependent – for example, final acceptance and end of defeats liability period. Broken lines indicate two events may have a relationship – for example, a variation order may affect time for completion or price or guarantees for performance.

What can be quickly seen is the extent to which the sectors are interrelated. Thus take over is significant in connection with:

- passing of guarantee tests
- reduction of liability for accidents, damage, and insurance
- release of part of the retention money
- possible commencement of the defects liability period
- application of any liquidated damages for delay
- unless property has passed before, property passes to the purchaser and risk in the works passes to the purchaser.

It is essential that this interdependence is borne in mind at all times when negotiating, drafting, or modifying forms of contract. It is so easy to alter or omit one clause without taking into account the consequential effects.

# General forms of contract

For engineering contracts within the UK and apart from the terms and conditions prepared by certain trade associations and major purchasers, the three most widely used conditions of contract are:

## Civil engineering
General conditions issued jointly by the Institute of Civil Engineers, the Association of Consulting Engineers, and the Federation of Civil Engineering Contractors, commonly known as the ICE Conditions. Current edition is the 6th. There is also a form with Contractor's Design.

## Building
Standard forms of Building Contract Sub-Contract and

Collateral Warranties prepared by the Joint Contracts Tribunal. The forms are known as JCT 80, IFC 84 for contracts of a lesser value and the Minor Works form. There is also a form for Design and Build.

### Supply and installation of mechanical and electrical plant
Form MF/1 which replaced the old Model Form A and for which a new edition was issued in 1995 known as MF/1 Rev. 3.

### Design, supply and construction of process plant
Model Form of Conditions of Contract for Process Plants issued by the Institution of Chemical Engineers.

### New Engineering Contract (NEC)
A new form of contract, which is in essence a set of core clauses to which can be added additional clauses for specific types of contract, has been developed under the aegis of the ICE.

It is intended for use on either civil, building or plant contracts and represents an important change from traditional forms.

In the international field the two standard forms most commonly used are those issued by the Fédération Internationale des Ingénieurs Conseils (FIDIC): one for civil engineering work and one for electrical and mechanical works.

However, in international contracting there is a far stronger tendency for individual employers to prepare and insist upon the use of their own forms of contract and in certain countries government departments and public authorities are required to do so by law. They may also be required by law to adopt certain standard tendering procedures.

In general these individual forms are tied in with the laws and legal system of the country concerned and impose upon the contractor a much greater share of the risks and responsibilities involved in the design and execution of the works. They are not intended to be fair or create a reasonable balance between employer and contractor, but rather to protect the employer's interests without much regard for those of the contractor.

Further, although such forms may refer to an 'engineer' it must not be assumed that his position is analogous to that of an engineer/architect under the terms of a UK contract, which have been defined judicially in the following terms:

the building owner and the contractor make their contract on the understanding that in all matters requiring professional skill the architect will act in a fair and unbiased manner and it must therefore be implicit in the owner's contract with the architect that he shall not only exercise due skill and care but also reach such decisions fairly holding the balance between his client and the contractor.

It must rather be accepted that he will consider his function to be that of protecting the employer's (and often his own) interests without any consideration for what is fair and reasonable. This point of difference is of crucial importance to the contractor when considering the reasonableness or otherwise of clauses such as those dealing with certification of payments, granting of extensions of time and determination of whether or not work is defective.

Examination of these forms shows that with certain variations one to another they all contain clauses dealing with the following points and generally in much the same way although certain clauses only appear in the export conditions. (This is not so true of the NEC and this form is the subject of a brief commentary on its own – see p. 189.)

Arbitration.
Assignment and sub-contracting.
Bankruptcy.
Certificates of engineer/architect.
Completion.
Contract price and terms of payment.
Contractor's default.
Contractor's equipment, vesting of.
Contractor's representatives and workmen.
Damage to property and injury to persons.

Defects liability.
Delivery of materials and passing of property.
Drawings.
Engineer/architect, appointment of, decisions of, representative of.
Execution of the work.
Faulty work.
Information.
Inspection and testing.
Insurance.
Language of the contract.
Law of the contract.
Patent rights.
Programme of work.
Provisional and prime cost sums.
Security for performance.
Site, possession of.
Statutory and other regulations.
Sufficiency of tender.
Suspension.
Termination.
Variations.

In the commentary on these clauses which follows they are examined objectively from both the contractor's and the employer's viewpoints. Where the clause is discussed at length in another chapter only brief reference is made.

### Assignment and sub-contracting
A distinction must be drawn both in the case of the purchaser and the contractor between the assignment of the benefit and that of the burden of the contract. In essence neither party can assign the burden of the contract, i.e. his performance obligations, without the consent of the other. This is implied by law and expressly provided for generally in the standard forms. As regards the benefit of the contract it is quite usual for the contractor to assign the right to receive payment so as to obtain funding for the contract, but generally under the standard

conditions the consent of the purchaser must be obtained. Whether or not the purchaser can assign the benefit depends on the terms of the contract and, in the current editions of the JCT and ICE forms, the consent of the contractor. If such consent is not obtained, any purported assignment would be void, both as regards breaches of contract which had occurred before the attempted assignment as well as those which arose afterwards (*Linden Gardens Trust Ltd v Lenesta Sludge Disposals Ltd*, 1993).

## Arbitration

The normal arbitration clause provides that any dispute arising under or out of the contract is to be referred to an arbitrator appointed in the absence of agreement between the parties by the president of some named independent authority or body.

However, certain foreign countries especially on government contracts or those with public bodies require that all disputes should be referrable either to their own Arbitration Tribunal or to the jurisdiction of their own courts.

The following points are relevant in considering whether or not to accept such a clause recognizing that non-acceptance may disqualify the tender:

1 How well developed is the legal system of the territory concerned and how independent in practice are the judiciary from the executive? How stable politically is the country, so that any answer given to those questions at the time of tender will be likely to remain valid during the course of the contract?

2 How tortuous, lengthy and costly are the legal procedures both for obtaining and enforcing judgement? Cases have been known to take at least six years.

3 Even if the employer, under pressure, were to agree to independent arbitration, would any award need to be enforced through the local courts of his country, because he has no assets elsewhere and, if so, do such courts have the power to re-open the award on the grounds of 'public interest'?

4 If the employer is a government department or government-

controlled agency, could they successfully plead sovereign immunity in any proceedings either to resolve the dispute or enforce an award?

In the UK the State Immunity Act 1978 broadly removes sovereign immunity from any government department or agency which enters into a commercial transaction, but no such situation necessarily exists abroad. Indeed it did not do so in the UK until the passing of the 1978 Act which reversed the previous law. If there is any doubt on this issue then the contractor should try to persuade the employer to agree to a waiver of immunity clause.

A case illustrative of the risks which may face a commercial firm in dealing with a State Trading Organization which is able to persuade its government to exercise sovereign powers on its behalf is that of *Settabello v Banco Totta and Acores* reported in the *Financial Times*, 21 June 1985. There the Court of Appeal refused to investigate the motives which had prompted the Portuguese Government to issue a decree law entitling Portuguese companies declared to be 'in a serious economic condition' to suspend the rights of their business partners to cancel contracts for non-delivery, although on the face of it the decree was solely intended to enable a particular shipyard to escape from suffering the penalty of non-compliance with its contract.

It may on occasions, particularly on a credit contract supported by ECGD, be possible to persuade the employer at least to allow any dispute relating to payment to be referred to ICC arbitration or to the courts of a neutral country, although in the latter case a check needs to be made that the courts would accept jurisdiction. More often than not, however, the contractor can only weigh up the risks involved in accepting the employer's clause and decide if these are sufficiently outweighed by the benefits to be gained from accepting the contract. What is essential is to make that judgement before tendering in full recognition of what the consequences of tendering will be.

Closely linked with the arbitration clause is that of the law of the contract (see p. 215).

Points which may need to be considered in the drafting or negotiation of an arbitration clause are as follows:

1   All disputes to be capable of being referred to arbitration. Attempts are sometimes made to provide that certain matters should be finally settled by one of the parties or someone acting on their behalf. This should always be resisted even as regards matters of fact.

2   The arbitration clause should be wide enough to allow matters to be reopened which during the period of the contract were decided by the engineer, or other person appointed to supervise the contract, according to his opinion. This could be done by including words such as 'The Arbitration shall have full power to open up, review or revise any decision, opinion, direction, certificate, or valuation of the engineer'.

   It is important that the clause is worded so widely since it has been held by the Court of Appeal in Hong Kong that an order issued by the engineer granting an extension of time was not 'a certificate' and therefore was not open to review in arbitration under the wording of the arbitration clause in the contract (*Costain International v Attorney General* [1983] 23 BLR 48).

3   The authority nominated to appoint the single arbitrator or umpire should be genuinely independent of both parties.

4   In home contracts there is no need to specify the arbitration tribunal or the rules under which the arbitration is to be carried out. With contracts involving a foreign buyer or seller, however, this should be done, and the fairest choice would seem to be the International Chamber of Commerce, see clause 49.3 of FIDIC (E & M) 1980. It should also be provided that the decision of the arbitrator(s) or umpire should be final.

5   In fairness to both parties the contract should provide that following the reference to arbitration and until the issue of the final award:

   ● the contractor should continue with the contract

- the employer should continue to make payments as they become due under the contract.

There is a tendency today to try and avoid either arbitration or action through the courts, by having the dispute settled by some form of alternative dispute resolution procedure. For example, the ICE 6th edition allows the parties to refer the dispute to a conciliator under the provisions of the ICE Conciliation Procedure. This procedure is not binding on the parties.

An alternative which is contained within many existing forms of sub-contract, the JCT 81 Design and Build and the NEC, is for adjudication whose decision will be binding on the parties unless either party refers it to arbitration. The primary distinction between adjudication and arbitration is that under adjudication the person appointed acts as an expert, is expected to use his own knowledge in deciding the dispute, and is not bound by rules of natural justice or evidence. He can act in an inquisitorial way as opposed to the adversarial-type procedure followed in arbitrations. Costs are usually borne equally by both parties. The outcome is not a judgement in the same way as an arbitration award and is only enforceable as a contract term, although it appears that it can form the grounds for a successful application for summary judgement under RSC 14.

Adjudication is intended to be quick and simple, perhaps even somewhat rough and ready. It is hoped that it continues this way and does not become hijacked by lawyers and turned into a process analogous to arbitration with lengthy submissions by counsel and evidence being given by expert witnesses.

### Bankruptcy and liquidation of the contractor
Under the circumstances the employer normally wants the option either to terminate the contract immediately or to give the receiver or liquidator the opportunity to complete the contract, subject to his giving appropriate guarantees. If there is any reasonable chance of the contractor being able to complete the contract the latter is normally the preferred step to take, since otherwise the employer faces all the delays and troubles involved in changing contractors, without much hope of

recovering his increased costs.

## Certificates

The contract will usually provide for certificates to be issued by the engineer or other supervising official in two different circumstances:

1 To record the date when some particular event occurred which is of contractual significance and to authorize the release of any retention moneys due at the point.
2 Only to authorize payment to be made to the contractor of the amount certified in the certificate as being then due.

Certificates falling under 1 are:

*Completion certificate* Issued normally under building or civil engineering contracts to record the date when work is substantially completed (see p. 257).

*Taking over certificate* Issued normally under plant contracts to record when the plant has passed its tests on completion (see further p. 257).

*Acceptance certificate* Used in process plant contracts to record the passing by the contractor of the performance tests.

*Final or maintenance certificate* Issued at the end of the defects liability period to record the end of that period. It may also depending on its working operate as a limitation on the contractor's liabilities under the contract – see Chapter 18.

Points to be noted in regard to the clauses providing for their issue are as follows:

1 The event giving rise to the right to claim the issue of the certificate should be clearly defined.
2 The certificate to record the date on which the contractor was entitled to claim its issue.
3 The certificate to be issued within a stated period of the date

on which an application is made which the contractor was entitled to make.

Certificates falling under 2 are usually referred to as interim or progress certificates. As they have no function other than to certify a sum of money for payment to the contractor, they have no contractual significance except for that purpose. Thus it is usually expressly stated that no interim certificate can be relied upon as conclusive evidence of any matter recorded in it and that the engineer can correct or modify anything in the certificate in any subsequent certificate.

## Contractor's default

As a weapon of last resort the employer must have the right to terminate the contract or take the work out of the contractor's hands, and either finish it himself or employ someone else to do so. In preparing this clause the draftsman must define:

1   The circumstances in which the employer's right to exercise this power arises.
2   The remedies which the employer has against the contractor on the exercise of such right.

No one can ever foresee all eventualities, so it is wise not to try to produce a comprehensive list of events entitling the employer to terminate, but rather to provide generally that he can do so should the contractor be in serious breach of contract and have failed to take any steps effectively to remedy the breach. The important safeguard here from the contractor's point of view is that the employer must first give notice of the breach complained of, and the period of such notice must be adequate to enable the contractor to take remedial action.

If the employer does take the work out of the contractor's hands, then the remedies which he has are normally the following:

1   To make use of all plant, material etc. on site for the purpose of completing the contract.

2   To retain any payments then due and not to make any further payments until the work is completed.

3   To apply any outstanding payments due to the contractor, and any outstanding portion of the contract price to the cost of completing the work.

4   If the costs of completing the work are greater than the balance of the contract price outstanding, then to recover the excess from the contractor.

5   Recover from the contractor the damages which the purchaser has suffered by reason of the contractor's default.

Sometimes a purchaser may in addition seek the right to recover from the contractor the money which he has paid the contractor for the part of the contract work already completed. A distinction must be drawn here between a default of the contractor which leaves the purchaser in possession of works or a plant of which he can make use after further work, i.e. have completed by another contractor, and a situation in which the works are useless to the purchaser, e.g. because the performance tests have shown them incapable, even after modification, of meeting the upper limit of the liquidated damages – see further, p. 330. In the former case it is reasonable that the contractor should retain the payments already made, subject to the purchaser's rights to damages. In the latter the purchaser has been deprived of the whole of the benefit of the contract. Effectively the only value which the works possess is as scrap material after the costs have been met of dismantling and reinstatement of the site. Under those circumstances the purchaser should have the right to reject and recover the interim payments already made.

## Contractor's equipment, vesting of

In order to provide the employer with additional security for interim payments which he makes during the course of the contract, if it usual to provide that the property in any constructional plant bought by the contractor on to the site vests in the employer until the contract is completed. The employer is then given the right to sell such plant should he be unable to obtain payment of any sums due to him.

It is important in drafting the clause to state expressly that the plant remains at the sole risk of the contractor who is responsible for any loss or damage to the plant in whatever way this is caused, other than through the fault of the employer.

One problem is that the plant may easily not be the property of the contractor but only hired, and in this case the clause would be inoperative as the contractor cannot pass to the employer the property in plant which he does not himself own, and the plant hirer is not of course a party to the contract.

The attempts made in the ICE 5th edition to deal with plant owned by others and hired to the contractor have been abandoned in the 6th edition. Clause 53 now deals only with plant which is owned by the contractor and does not therefore cover plant which is the property of third parties.

## Contractor's representative and workmen

The employer is concerned to ensure that:

1   The contractor has a competent representative on site during the time work is proceeding.
2   He can require the contractor to remove from any site any person to whom the employer objects on the grounds of negligence, incompetence or undesirable conduct.
3   The contractor does not recruit his labour from the employer's own workpeople without the employer's consent.
4   The contractor provides all necessary facilities – for example, canteen, first-aid – unless the employer is prepared to allow use to be made of his own facilities.
5   The contractor complies with all relevant laws, regulations and customs as they affect his workpeople. Note that in the ICE 6th edition, clause 34, which dealt with the implementation of the Fair Wages Resolution of 14 October 1946, has been dropped and the clause is no longer included in the contract conditions.
6   On overseas contracts the contractor will also be concerned

with the right to bring in labour, the issue of work permits and visas and the time when these will be made available. Security clearances may need to be obtained and these usually involve the employer in sponsoring the employee. Although the employer cannot be expected to undertake that work permits etc. will be issued, since this is not his responsibility, the contract should at least provide that the employer will use his best endeavours and provide every assistance.

## Contract price and terms of payment
The methods of determining the contract price are described in Chapter 10, and the definition of the price should be included in the letter of acceptance or contract agreement (see p. 165).

Terms of payment are dealt with in Chapter 14.

## Damage to property and injury to persons
For the detail on this, including the different approach adopted by each of three sets of standard conditions referred to on p. 166, see Chapter 19.

## Defects liability
See Chapter 18.

## Delivery of materials and passing of property
See Chapter 17.

## Drawings
If the contractor is responsible for design, the contract will usually require him to submit drawings of the works for the approval of the purchaser or his engineer. On completion of the contract the contractor will be required to provide a set of the drawings necessary for the operation and maintenance of the works for the purchaser's use. Points to be noted in connection with these requirements are:

1   Approval of the drawings. There should be a specified time limit for approval, and if no comments are received within so

many days then the drawings should be deemed to be approved. Delay in the approval of drawings is a frequent cause of delay in the completion of the contract.

2   Drawings to be submitted should not include shop or fabrication drawings, as these are rightly regarded as confidential to the contractor.

3   The as-built drawing of the works which are supplied to the purchaser for the purpose of operation and maintenance of the works should remain the property of the contractor and not be used by the purchaser for any other purpose. By buying the plant the purchaser does not buy with it the designs or drawings so as to enable him to use these for other purposes, for example extending the works or pirating spares the design of which is the copyright of the contractor.

This position is not affected by the decision in *British Leyland v Armstrong Patents* so long as the designs are the subject of protection either by patent rights or registered designs. What that case did establish was that in the absence of such protection a manufacturer could not use his copyright in drawings to prevent a purchaser from having spare parts manufactured by a 'pirate'. Nor can a manufacturer seek to include in his conditions of sale a requirement that the purchaser shall obtain such spares only for him.

If the purchaser wants the ownership of the drawings then this should be the subject of separate arrangement, if in a particular case the contractor is prepared to agree to this. Normally he would only do so under a form of licence agreement which would provide for further payments to be made, either in the form of a once-and-for-all lump sum or more likely proportionate to the further use made by the purchaser of the design rights. The selling price would then reflect the contractor's accumulated know-how and design effort which has gone into the development of the designs.

4   In the same way the contractor must keep confidential and not make use of or disclose drawings and information supplied by the purchaser except in so far as it is necessary for him to do so for the purpose of carrying out the contract.

5 The as-built drawings will be required by the employer at the time when he starts to train his personnel to operate the plant and takes over responsibility for maintenance. The contract should provide therefore:

- the numbers of copies, and form of the reproducibles, velographs or micro-films
- the programme for their handing over to the employer, recognizing that provisional copies only may be available initially and that final copies will have to follow after hand over of the plant.

# New Engineering Contract (NEC)

The New Engineering Contract (the NEC) is a fundamental departure from the traditional forms of contract. The intention was to make improvements under three main headings:

- Flexibility, so that it can be used for any or all of the traditional disciplines
- clarity and simplicity, so that it is written in plain English and not legal language
- to act as a stimulus to good management.

The NEC exists in nine sections:

1 General.
2 The conductor's main responsibilities.
3 Time.
4 Testing and defects.
5 Payment.
6 Compensation events.
7 Title.
8 Risks and insurance.
9 Disputes and termination.

Within each section there are the core clauses, which will remain unchanged irrespective of which price option is selected as described below.

For each section there are then the main option clauses. These offer the choice of a different basic allocation of risk between the employer and the contractor according to the method of pricing used:

Options *A* and *B* are price contracts, *A* with an activity schedule and *B* using bill of quantities.
Options *C* and *D* are target contracts in which the financial risks are shared by the employer and contractor in agreed proportions.
Options *E* and *F*; *E* is a form of cost reimbursable contract and *F* a form of management contract.

There are then secondary options which may, apart from a few instances, be used with any of the main options, covering:

performance bond
parent company guarantee
advanced payment to the contractor
multiple currencies (Options A and B only)
sectional completion
limitation of the contractor's liability for his design to
    reasonable skill and care
price adjustment for inflation (not used with *E* and *F*)
retention (not used with *F*)
bonus for early completion
delay damages
low performance damages
changes in the law
special conditions of contract (only to be used exceptionally).

An NEC contract therefore comprises:

the core clauses
the main option classes applicable to the method of

procurement chosen, and
the secondary options selected by the purchaser.

Note that the contract is not related specifically to the type of work. The same form can be used for building, civil engineering or plant design and construction. It is also the only form which can be used for multidisciplinary contracts.

The following lists some of the main features of the NEC and the way in which it operates:

1   Essential to the NEC is the works information which describes in detail what the contractor has to do and will include a number of matters which in conventional forms would appear elsewhere in the contract. It will include, for example, the extent of the contractor's responsibility for design. It is assumed that the contractor will always have some design responsibility.

2   The traditional role of the engineer/architect is divided into four:
   - project manager
   - designer
   - supervisor of construction
   - adjudicator of disputes.

   The first three functions are carried out on behalf of the employer. The fourth is carried out independently. It follows that while it would be possible for the first three roles to be performed by the same person, although this is not recommended, the adjudicator must be a different person.

3   The intention is to reduce to a minimum the amount of design work to be done post contract. Where the option of a firm price is chosen the information provided at time of tender is required to be sufficient to enable the works to be constructed without further instruction.

4   There is no provision for nominated sub-contractors/suppliers.

5   The contract provisions are designed to allocate risks sensibly between the parties and in a way which is intended to encourage good management.

6  Emphasis is placed on the planning and programming of the work monitored by the project manager.

7  The use of a system of identified compensation events which cover situations in which the contractor would expect remedies in terms of cost or time, e.g. variations, late instructions etc. The contractor is required to give a quotation showing the effect of the event which is assessed by the project manager. For compensating events entitling additional payment the assessments are based on actual costs incurred not the rates and prices in the contract and when assessing time extensions no account is to be taken of the contractor's float.

8  The introduction of the adjudicator, whose identity must be specified by the contract date and to whom all disputes must be referred for resolution with the right for either party, if dissatisfied, to refer the matter to arbitration.

The use of the NEC is growing and the first experiences have been extremely favourable. However, it is emphasized that this is not just a new set of conditions but a new way of thinking about construction contracts. The thrust of the NEC is towards better management and a dramatic reduction in the adversarialism which has plagued UK construction sites for years. To apply the NEC successfully needs therefore a radical change of approach by all involved.

Sir Michael Latham in his report recommended certain amendments to the NEC but also strongly recommended its adoption as amended in both the public and private sectors.

# 12 Standard terms of contract: II

The position and authority of the architect/engineer under certain forms of contract developed within the UK are dealt with in Chapter 21. The relationship between the engineer and the purchaser and the contractor as described in that chapter is, however, largely unique to those forms of contract and to international forms such as those prepared by the Federation of International Consulting Engineers which are largely based on UK practice. It is important, when preparing contract conditions, to be clear as to whether the person or body exercising supervising powers on the purchaser's behalf will be acting in that role or not. It they are, then it may be appropriate to invest them with wide discretionary powers, for instance in the pricing of variations or the granting of extensions of time. If, however, it is known that they are merely acting as agents on the purchaser's behalf and have no capability for the exercise of independent professional judgement, then the contract should be drafted so as either to leave matters to be agreed between purchaser and contractor (pricing of variations) or to give the party concerned an absolute contractual right in certain events which is not dependent upon discretionary judgement (extensions of time).

## Architect/Engineer – decisions and instructions

Where the architect/engineer is acting in his independent professional role then the contract will normally provide:

1  That the contractor must comply with all decisions of the architect/engineer, subject only to the contractor's right to challenge such decisions at arbitration. This proviso is important, as there would appear to be no implied obligation on the part of the architect/engineer only to issue reasonable instructions and, if the contractor has bound himself to comply with any instructions the architect/engineer may issue, he could find himself in some difficulty if the architect/engineer concerned was being awkward.

2  That after acceptance of the tender decisions and instructions will be given only by the engineer.

By the inclusion of this second provision in the contract conditions the employer has disqualified himself, *vis-à-vis* the contractor, from interfering in the administration or control of the contract, although, as explained on p. 372, this does not prevent him from giving instructions to the engineer, provided that these do not improperly restrict the exercise by the engineer of his discretionary function. The contractor for his part has accepted that he must take instructions only from the architect/engineer.

If the engineer is an outside consultant it requires the exercise of great discipline and self-restraint on the part of the employer, particularly an employer who has engineering and contracts departments of his own, to make these particular provisions operate successfully. Too often the employer's own staff will start giving instructions to the contractor direct. Unfortunately also from the contractor's point of view, if a dispute does arise later on, for example as to whether he is entitled to be paid for some extra work done on the verbal authority of someone other than the engineer, then the contractor may well find the terms of the contract quoted against him and his claim disallowed. This of course applies only in so far as the engineer has authority under the contract; it would not apply to a change in the contract conditions since the engineer has no authority to make such changes.

The moral is again that the terms of the contract must be related to practicalities. If the employer wants his own engineering department or project manager to have the last word and be able to deal with the contractor direct, then he should never have appointed the consultant as engineer under the contract. If in order to maintain progress, or because of the isolation of the site, it is necessary for people other than 'the engineer' to give instructions for extra work, say up to a certain financial limit, then this should be provided for specifically in administration procedures issued by 'the engineer' and circulated to all concerned, including the contractor.

### Execution of the work

The contract will normally provide that all work must be executed in accordance with the manner set out in the specification, or where not so set out to the reasonable satisfaction of the engineer. From the contractor's viewpoint it is important that the word 'reasonable' is included in the clause to make it clear that the engineer is required to act in a reasonable manner, and to ensure that the contractor has the right to challenge the decision of the engineer at arbitration if he considers that the engineer has acted unreasonably. It is prudent to couple this with wording in the arbitration clause which emphasizes the point (see p. 181).

It is important to note that where the contract provides that work is to be performed in accordance with the specification *and* to the reasonable satisfaction of the engineer these words are likely to be treated as creating two separate obligations. Accordingly it follows that a certificate issued by the engineer is not necessarily conclusive that work has been executed in accordance with the contract – see *National Coal Board v William Neil & Son* [1984] 1 A11 ER 555, where a decision to that effect was reached on the wording of the BEAMA Standard Conditions RC version, January 1956 edition. It is important to note however that each such case will be decided in accordance with the wording of the particular clause and the facts of the individual case and that there are authorities which point in the other direction.

### Faulty work

If during the course of the contract the engineer decides that any work carried out by the contractor is defective or does not comply with the contract, then it is usual to provide that he can require the contractor to correct this and, if necessary, re-execute the work or take away the defective items and replace them with ones which do comply with the contract.

This should be sufficient protection for the employer, but just in case the contractor fails to replace defective work, MF/1 goes on to give the employer the right to do the work himself and charge the contractor with the additional costs incurred, provided that these are reasonable (see clause 26). It also expressly provides that any action taken by the employer under that clause will not affect the employer's right to claim damages for delay, so that not only must the contractor pay the costs of putting the work right, but he also almost certainly faces the prospect of paying liquidated damages when the works are finally completed.

Similar provisions are contained in other forms of contract, e.g. the ICE conditions, clause 39(2). That clause, unlike clause 26 of the MF/1 conditions, contains an express provision that the engineer may deduct the amount of the costs incurred by the employer from any moneys due or which may become due to the contractor. It is considered however that under the general principles of law the engineer under an MF/1 contract could deduct the amount of such additional costs from any future certificate but he would not have power, as he does not under the ICE form, to withhold a certificate in order to ensure that sufficient funds were available, nor would the employer have the right to make a deduction from moneys due under any other contract between the employer and the contractor unless there was an express agreement between the employer and the contractor to this effect.

## Mistakes in information

Where information is to be provided by one party to the other for use in connection with the contract, or other work which the purchaser may be carrying out as part of a project – for example, plant positions and loads which the purchaser requires for foundation design – the contract usually states that the party providing such information is responsible for any errors which it contains, and for meeting the additional costs caused by such errors. On the face of it this is a reasonable provision, but if applied too rigidly in practice it can cause difficulties.

Both parties are usually pressing the other for drawings and information. The earlier these are released the greater the probability that they will contain errors, or at least that the party supplying them will want to make changes to them as his own design develops. It may be reasonable, therefore, initially to release data or drawings which are marked 'provisional' and for which contractual responsibility is not accepted, and to follow these up at a later date with final issues to which the provisions of the clause would apply.

In the clause in MF/1 it appears arguable that the contractor's liability is limited to the cost of alterations or remedial work *to his own work* due to the errors in the drawings. But the costs which the purchaser is most likely to suffer are the additional costs he will have to pay to others, e.g. the civil contractor who has now to re-execute his foundations. It is suggested that the original Model Form MF/1 wording 'or if the same be done by or on behalf of the Purchaser shall bear all costs reasonably incurred therein' makes it clearer that the contractor's liability extends to re-work which has to be done by others.

## Inspection and testing

No contractor can reasonably object to the purchaser or someone on his behalf having the right to inspect and test work which is in progress or which has been completed in the contractor's works or in the works of his major sub-contractors. But inspecting and testing provisions should be fair to the parties and practical in relation to the circumstances of the particular contract, and the following points need particular consideration:

1   The extent to which the inspecting authority is given the right arbitrarily to reject. If an inspector rejects work, he should be required to state the reasons for such rejection in writing and his decision should be subject to challenge at arbitration. The inspector's decision should never be final and binding.

2   Most companies have their own internal inspection and testing procedures, the costs of complying with which are allowed for as part of normal selling costs. If the employer wishes to impose special testing procedures, these should be clearly defined in the specification, so that the contractor has the opportunity of allowing for the costs of these when tendering, and possibly of indicating to the purchaser the reduction in costs which would be possible were he to dispense with these procedures.

One particular provision of which the contractor must be especially wary is that which allows the employer or the engineer to add tests additional to those included within the specification. Such a provision has in the author's experience been used deliberately by an overseas purchaser to force a contractor into delay so that the purchaser would exact a penalty! Admittedly this is an extreme case but even without such intent the use of such a provision by the type of young, over-zealous, academic engineer so often found in developing countries can have disastrous effects on both a contractor's costs and programme and its inclusion should be strongly resisted. If the right to add additional tests must be conceded, they should be limited to those of a similar nature to the ones specified in the contract, e.g. exclude 'type' testing, and there should be equally a right for the contractor to an extension of time and the payment of additional costs.

3   Inspection and testing takes time, and if the purchaser is calling on the one hand for an extremely tight delivery schedule he cannot on the other hand expect to be able to insert into the manufacturing programme his own special inspection and testing requirements. This is particularly the case if to comply with these will mean putting a hold on the

manufacturing programme at varying intervals during its execution. One essential proviso in any event is that if at the time appointed for carrying out tests, or if after reasonable notice has been given, the inspector fails to attend the tests, the contractor can proceed in his absence.

4    If the plant fails to pass the tests, any repeat tests must be carried out at the contractor's expense.

## Insurance
See Chapter 19.

## Patent rights
A patentee who believes his rights have been infringed would in most instances proceed against the person who he claims is making use of his patent rights rather than against the contractor who built or supplied the plant concerned. The purchaser wants to be certain, therefore, that his use of the works is not going to be interfered with in the future by someone claiming that it is an infringement of his patent rights, or that , if this does happen, he has the right of indemnity against the contractor. This is reasonable provided that:

1    The infringement is not due to the contractor having followed a design or instruction given by the purchaser.
2    The purchaser is not making use of the plant in some way which is different from that indicated to the contractor or reasonably to be inferred by the contractor at the time of entering into the contract. This would apply particularly to process plant where a patent may relate to particular temperature or pressure conditions or operation in a particular manner.

Equally the contractor for his part wants an indemnity against his infringing any patent rights through following designs or instructions which he is given by the purchaser.

## Programme of work
All construction contracts require the contractor to produce a

programme showing the order in which he proposes to carry out the works. A preliminary outline of such a programme is usually included with the tender. On simple projects this will be in the form of a bar chart; on more complex projects it will be in the form of a network analysis.

The main issue which arises is the contractual status of such a programme. Is it part of the contract so that the contractor is contractually bound to complete the various operations shown on the programme by the stated dates and equally is the employer contractually bound to allow him the facilities to do so? Alternatively is the programme only a representation of the manner in which the contractor intends to proceed so that provided he meets the contractual date for completion there can be no claim against him by the employer if certain of the intermediate dates given on the programme are not achieved? Under those circumstances the contractor would only be in breach of contract if completion of the operation by that date had been specifically made a contractual obligation, e.g. the date was a sectional completion date.

Unless the contract specifically provides otherwise (and none of the standard forms referred to on p. 175 does), it is considered that the programme is not a contractual document in the sense that the contractor would be in breach of contract if he failed to meet one of the intermediate dates for completion of a particular operation. If however the contractor were to fail significantly in meeting an intermediate date on the critical path or it was evident that he would do so unless corrective measures were taken, then this would in the author's view be grounds for action by the employer/engineer under, for example, clause 46 of the ICE conditions or its equivalent. It could also, depending upon the seriousness of the delay, constitute an anticipatory breach of contract.

However it is to be noted that clauses such as that do not impose a positive obligation on the contractor to proceed to execute the works with due diligence and expedition, but provide a remedy should he fail to do so – per Staughton J in *GLC v Cleveland Bridge & Engineering Co. Ltd.* In any event the meaning of the words 'due diligence and expedition' must be interpreted in the light of the contractor's other obligations as to time under the contract and their true meaning is 'with such diligence and expedition as were reasonably required to met the completion date in the contract' per Staughton J in the same case and confirmed on appeal – 8 CON LR p. 30.

Similar wording in clause 21(1) of the JCT 1963 conditions (repeated in 23.1 of JCT 80) does require that the contractor should proceed 'regularly and diligently' with the works and that was considered by the Court of Appeal in *West Faulkner Associates v Newham London Borough Council*, reported in *The Times* 18 November 1994. In summary the contractor's obligations were said by the court to be 'to proceed continuously, industriously and efficiently with appropriate physical resources so as to progress the works steadily towards completion substantially in accordance with the contractual requirements as to time, sequence and quality of work'. In that case it was clear that the contractors were proceeding regularly (in the sense that they had sufficient men, materials and plant on site) but not diligently, in that the resources were not being managed and applied industriously and effectively. One was no use without the other.

As regards the position where the programme shows a completion date earlier than that contained in the contract, it has been decided in a case on the JCT 63 form clause 21 (1) that:

- the contractor was entitled to carry out his work in accordance with the accelerated programme, but
- the employer and his agents only have a duty to do that which is reasonably necessary to enable the contractor to comply with his obligations. Since the contractors had the right, but not the obligation, to finish early, the employer and his agents accordingly had no duty to provide him with instructions at such times as were necessary to enable him to achieve the earlier completion (*Glenlion Construction Ltd v The Guinness Trust*, 11 CON LR).

Both for the purpose of good contract administration and in order to safeguard his interests the contractor should supplement the programme with an appropriate procedure and notices to the engineer of his requirements for drawings and information as the programme on its own may not be sufficiently detailed.

It equally follows that if the employer were similarly to fail to provide facilities which contractually are his responsibility by the programme date or it appears evident that he will do so, then the contractor could require the employer to take appropriate measures to speed up the work in question or provide the contractor with compensation. The programme in effect puts the employer on notice as to the contractor's requirements if the contractor is to satisfy the completion date, and the employer is then bound to meet these or he must both grant the contractor an extension of time and meet the additional costs directly and necessarily incurred as a result of his default.

If therefore the employer wants to make intermediate dates, e.g. for the supply of drawings or access to buildings contractual, so that he can claim damages against the contractor for failing to meet these, then he must say so specifically in the invitation to tender or at the stage of contract negotiation since it will affect both the contractor's assessment of his risks and also possibly the order and method in which he intended to carry out the works. Certainly it will restrict the contractor's flexibility of operations which could have a major cost impact.

On large, complex, multidisciplinary plants there is an argument for the employer making intermediate dates contractual but he must recognize that it will increase the price, lead to a rigidity of attitudes and lack of give-and-take in site working and to an increase in the unproductive paper war of claims and counter-claims (see Chapter 15).

### Provisional and prime cost sums
See Chapter 14.

### Security for performance
There are three types of bond or bank guarantee which the contractor may be required to provide:

● advance payment bond

- contract performance bond
- maintenance or retention bond.

Before discussing each of these it is necessary to distinguish between two types of bond/guarantee.

The first type is what is known as a conditional bond, that is the rights of the employer to call upon the bond is dependent upon some trigger event having occurred which is related to the non-performance of the contract. Unfortunately the nature of the bond has been hidden by some obscure and archaic form of drafting which has caused the courts, and the parties, considerable problems.

The usual form of drafting has been to begin the bond with words which bind the bondsman and the contractor to pay the employer a certain sum of money, and then to provide that if certain conditions are satisfied the obligation to pay is null and void.

Recent case law has demonstrated a number of times the weaknesses of this form of drafting. It is not stated when a claim can be made under the bond, nor if the damages claimed have to be proved, nor whether the bondsman can set off claims which the contractor has against the employer. One of the difficulties is whether or not the bond constitutes in law a 'guarantee'. In the previous edition reference was made to the case of *Bold and Another v BGK Metals and Another* 1987, in which it was decided with some hesitation, and perhaps with a desire to see justice done, that the form of bond used there was indeed a 'guarantee'. However, in the recent case of *Trafalgar House Construction (Regions) Ltd v General Surety and Guarantee Co.* the Court of Appeal decided that the usual form of conditional bond is not a 'guarantee' since there is no obligation on the bondsman to 'see to it' that the guaranteed obligations will be performed, but only an obligation to pay up to the stated amount of the bond.

The practical effect of this is that the bondsman cannot exercise any right of set-off or counter-claim which belongs to the contractor.

Further, according to the Court of Appeal, there is no obligation on the employer to prove default of the contractor – it is only necessary for some default to be shown. Nor is it necessary for the employer to prove the damages he has suffered. All he needs to do is to show that he has made a bona fide estimate of his damages.

Obviously the same position applies as between a main contractor for whose benefit a sub-contractor provides a bond.

The decision of the Court of Appeal has caused a considerable stir, since it seriously weakens the defences open to a bondsman and exposes therefore the contractor or sub-contractor, who has only obtained the bond by providing the bondsman with a counter-guarantee, to a far greater risk than had previously been thought to be the case. At the time of writing it is understood that an appeal is to be made to the House of Lords.

Since the form of the bond in the *Trafalgar House* case was the archaic one in general use with UK construction forms of contract, the decision – apart from causing consternation amongst contractors – may at last lead to the abandonment of the form only some sixty years from the time when it was strongly condemned by the House of Lords.

The alternative form of bond is usually referred to as a 'cash' or 'on-demand' bond. As its name implies the bond can be called by the employer on first demand *and without having to prove any default on the part of the contractor*. If such a bond is called by the employer then in so far as it is unconditional the bank must pay and will then look to the contractor for reimbursement of the money under the terms of the agreement under which it provided the bond. At one time such bonds only tended to be required by overseas purchasers, especially in the Middle East. Today the position has changed and many UK and continental purchasers insist on bonds being essentially in an 'on-demand' form.

The position of a UK bank which gives an unconditional on-demand bond has been stated by Lord Denning in these words:

A bank which gives a performance guarantee must honour that guarantee according to its terms. It is not concerned in the least with the relations between the supplier and the customer; nor with the question whether the supplier has performed his contracted obligations or not; nor with the question whether the supplier is in default or not. The bank must pay according to its guarantee on demand, if so stipulated without proof or conditions. The only exception is when there is clear fraud of which the bank has notice.

The question of fraud was considered in *United Trading Corporation & Others v Allied Arab Bank Limited & Others*, FT Commercial Law Reports 17 July 1984. There the Court of Appeal stated that the sellers could obtain an injunction restraining a bank from paying out on an on-demand performance bond but that there must be clear evidence of the fraud of which the bank had knowledge. It was stated that 'if the court considered that on the material before it the only realistic inference to be drawn was that of fraud then the seller would have made out a sufficient case of fraud'. Unfortunately for the seller the foreign buyer, an Iraqi state organization, refused to submit to UK jurisdiction, which the court held to be reasonable in all the circumstances, and there was therefore no opportunity for the seller to inquire into the honesty or otherwise of the buyer's belief in the validity of his claim to call the bond.

It is obvious therefore that in the hands of an unscrupulous employer, especially if the contract is subject to the jurisdiction of a territory whose legal system provides the English contractor with no effective remedy, the use of 'on-demand' bonds can be abused. In the UK and the continent there is the protection that the contractor, if he considered the call on the bond to be unjustified, could proceed against the employer with the expectation of securing the repayment of any sums which he was not liable to pay under the terms of the construction contract, although this could be a lengthy process.

At the same time 'on-demand' bonds do have distinct advantages to both the purchaser and the bank. The purchaser does not have to establish loss or breach of contract before he can obtain his money. The bank is not involved in any disputes as between the contractor and the employer. The bank is only concerned with whether or not the contractor has sufficient funds with which to meet his counter-indemnity to the bank. For this reason banks exercise considerable caution in giving such bonds and take their amount into account when deciding on the level of the contractor's borrowing facilities.

Assuming the *Trafalgar House* decision is upheld however, there are two factors present in the UK which will reduce the difference between the two types of bond. First, with a conditional bond the employer may be able to obtain immediate access to funds merely by alleging default and submitting a bona fide claim for damages without having to prove either of these. That he should be able to do this was the purposive basis upon which the Court of Appeal reached their decision. Secondly, the banks allow that the 'on-demand' bond may be worded in such a way that the employer must provide a statement as to respects in which the contractor is in default and that he is due to the sums claimed. But the banks will not go behind such statements and will accept them on their face value. Only if the abuse is apparent on the document itself will a bank decline to pay.

In certain other parts of the world, especially the Middle East, there are other problems which may arise with such bonds:

1    Even if the bond is time-expired it may not be treated by the bank as being no longer enforceable unless it is actually physically returned, and securing the actual release of the bond in certain territories is extremely difficult. Local banks in the Middle East are particularly vulnerable to pressure from government authorities and have indicated that they would pay on a bond, even if it were time-expired, and there is every reason to believe that in turn the UK bank would pay and then recover from its client.

The UK bank may be prepared to accept written notice from the contractor that the bond has time-expired as evidence that any call must be fraudulent, but it is more probable that the bank would refer the matter to the local bank before doing so, who would in turn refer to their customer and that could well precipitate the bond being called!

2    So long as the bond is outstanding it will be treated as part of the contractor's contingent liabilities and therefore against his overdraft limit. Having regard to the value of overseas

contracts and their total bonding requirements, this can impose restrictions on the amount of business for which the contractor can bid and undertake.

Assurance must be obtained from the bank or banks before submitting a bid that if it is successful bonding facilities will be available since otherwise the contractor may find himself in the position of not being able to sign a contract and so lose his tender bond.

Because of the problems associated with bonding requirements and the risk of unfair calling ECGD provides certain assistance to companies tendering overseas. Cover can be obtained against unfair calling and there is also the facility of bond support which allows the bank to recover directly from ECGD in the event of a call and who in turn then recover from the contractor unless the call had been made 'unfairly'. The advantage to the contractor of this bond support is that the bank will generally be prepared to provide bonds which are 'bond-supported' to an unlimited extent.

In order however to obtain either of these facilities from ECGD the contractor must have taken out basic ECGD cover and the total cost of this plus the bond support facilities will often be quite significant in the highly competitive market for overseas constructional work. For many companies however there is just no alternative given the risks otherwise involved. The problem of bonding in relation to cash flow and terms of payment is considered further in Chapter 14.

It may be suggested that one way of reducing these risks would be for the contractor to include safeguards in his tender/contract. The problem with including them in the tender is that in the buyer's market which generally prevails in overseas construction there is every chance of a qualified tender being summarily rejected. Only if the contract is on a negotiated basis and the contractor has some strength in his negotiating position, is it considered feasible for him to seek a modification of the on-demand requirement. He will also have to recognize that he is unlikely to be supported by his bank since UK banks prefer a clean situation in which they avoid any involvement in the relations between the contractor and the employer, even though the contractor is their client.

If however it does appear there is room for some negotiation, the following suggestions are made:

1   The contract could be worded so that the bond only applied to sums which were 'properly' due. This at least establishes a negotiating platform.
2   Before being entitled to draw any sums from the bond the purchaser to be required to give the contractor reasonable notice of his intentions.
3   The purchaser should only be entitled to draw from the bond an amount equal to the damages or costs which he claims are due.
4   Before obtaining payment the purchaser should have to produce a copy of his warning notice to the contractor, a statement setting out the contractor's default, and an invoice for the costs or damages due.
5   The bond should, again ideally, only be payable in the UK.
6   The validity of the bond should expire at the end of the defects liability period or not later than a fixed date which should allow for a reasonable extension of the defects period.
7   While from the purchaser's viewpoint a provision that if he makes a call on the bond he can require the contractor to reinstate the bond to its full value is ideal, it places the contractor in a very vulnerable situation if the right is unlimited. There should at least be a limit on the total amount to be bonded, say of 30 per cent of the contract price, if the bond itself is for 10 per cent.

With a reasonable purchaser these precautions should be unnecessary. Unfortunately recent commercial history has shown that not all purchasers, particularly certain overseas organizations, can be relied on to be reasonable, and the threat of cashing the bond has been used as a potent weapon of blackmail, more especially in circumstances where the right to go to arbitration has been of limited value due to the lack of genuine independence of the arbitration tribunal.

## Parent company guarantee

Where the contractor is a subsidiary of a larger firm or of a group it is essential in the purchaser's interests that he obtains a guarantee from the parent company of the performance by the subsidiary of its obligations. The subsidiary is unlikely itself to own any assets and in the absence of such a guarantee the parent would not be liable for the subsidiary's default. The guarantee should be unlimited in amount and should require the parent to perform or have performed the subsidiary's obligations. But the parent's liability should be expressed to be no greater than that of the subsidiary, so that the parent could set up as against the purchaser any defence open to the subsidiary.

## Site, possession of

The contractor must obviously be given possession of the site to enable him to perform his contract; indeed this term is implied by law. But such possession is not normally exclusive. The employer will want access; so may other contractors. If, however, there are any serious restrictions on the availability of the site or the operations of several contractors have to be dovetailed together in a limited space, this should be set out expressly in the invitation to tender. This applies particularly to contracts for repairs or extensions to existing buildings which must continue in use while the work is being carried out.

This latter point is significant in avoiding claims since in the absence of any specific provisions in the contract to the contrary it will be held that there is an implied term in any construction contract that the contractor will be given sufficient, uninterrupted and exclusive possession of the site as will enable him to carry out his work unimpeded and in the manner of his choice. A general clause providing that no implied obligations were to be included in the contract would not be sufficient (see *The Queen in Right of Canada v Walter Cabott Construction Co.* [1975] 21 BLR 42).

## Statutory and other regulations

The contract should be carried out in compliance with the laws and regulations applicable to the works. If these restrict the methods of working or the use of certain materials, or prescribe the way in which plant has to be designed, then they are all factors the contractor must take into account in pricing his tender. It is his duty to do this irrespective of whether the employer has expressly drawn his attention to these requirements or not, provided:

1   The regulations are not purely internal safety rules of the employer of which the contractor was unaware.
2   Where the regulation is only broken by the use of the item in a particular manner or place, if the contractor was either expressly or by implication made aware of the use to which the item was to be put.

What, however, if the regulations are changed part way through the contract? Obviously the contractor must still comply, but if doing so costs him extra money he ought to be entitled to recover these costs from the employer (see, for example, MF/1, clause 6.1).

Although when operating overseas the contractor must again comply with all local laws, statutory instruments and regulations as they affect the carrying out of the contract, it is essential for him to include specific provisions which entitle him to additional costs should these be changed during the course of the contract, e.g. an increase in national insurance charges or the requirement that charges of this type be applied to ex-patriate as well as local employees. An example of such a clause is set out below.

If after the date of Tender any of the events listed in sub-clause (2) below shall occur and such event results in an increase or decrease in the cost to the contractor or any sub-contractor to the contractor of the carrying out of the Works, then the amount of such increase or decrease shall be added to or deducted from the contract price

(2)   The events referred to in sub-clause (2) above are:
   (a)   the introduction of any new . . . . . . . .[insert name of the territory] legislation
   (b)   the amendment of any existing . . . . . . . . . . . legislation
   (c)   any change by the appropriate . . . . . . . . authorities in their

interpretation of any existing . . . . . . . . . . . . . legislation
(3)  For the purpose of the above sub-clauses the term 'legislation' shall be construed in its widest sense and shall include any enactment or decree or any form of subsidiary regulation or legislation duly enacted by a competent authority.

A further point which arises is in relation to codes of practice or recommendations of such bodies as the CCITT in the international telecommunications industry. It is essential that the contractor's tender and the contract are tied to such codes of practice or recommendations so far as they have been published at the date of his tender. They should therefore be unambiguously identified in the contract documents. If this is not done then the employer may seek to argue that it is the latest codes or recommendations in force at the time of completion which should apply. This is not theoretical; the author has personal experience of such a claim being advanced.

## Sufficiency of tender

It is usual to include in the contract conditions a provision that in tendering the contractor has taken all risks and eventualities into account which may affect his tender price, so that he cannot afterwards put forward a claim based on lack of knowledge as regards the site, conditions under which the work is to be carried out, and so on (see, for instance, clause 11(2) of the ICE conditions). Particularly, however with civil engineering work, it would be unreasonable to make this an absolute obligation, since one can never rule out the possibility that the information available at tender stage may give the wrong picture of the conditions which will be encountered – for example, boreholes which happen to miss large boulders in otherwise soft ground. Indeed in clause 11(2) it is stated that the contractor shall be deemed to have satisfied himself 'so far as is practicable and reasonable'.

Normally, therefore, the contract also provides that if the contractor encounters physical conditions or artificial obstructions which could not reasonably have been foreseen by an experienced contractor, then the contractor is entitled to the reasonable additional costs of dealing with these conditions (see clause 12(1) of the ICE conditions). This particular clause has been the subject of frequent and costly disputes between the parties. In practical terms tendering periods do not allow the contractor any opportunity to do other than inspect the surface of the site and examine the data provided by the employer. If this is inadequate, as is frequently the case, claims by the contractor under clause 12 are almost bound to arise. Since, if such a claim succeeds, the contractor is paid the reasonable costs which he incurs in overcoming the problem plus an allowance for profit, it is not surprising that contractors frequently view this clause as providing a means of increasing their margin on the contract.

One solution to this problem would appear to be for the employer to carry out a much more detailed site and sub-soil investigation pre-tender, for this information to be made fully available to the tenderers, for the tenderer to be required to take the risk of ground conditions and to be allowed to price this risk into the contract price. This would only seem to be feasible, however, if the contract were on a design and construct basis.

A modification to this would be to require the tenderers to undertake the ground investigation themselves and to be allowed to employ a single consultant for this purpose on a 'pooled' basis. This approach has been used by the Scottish Office Roads Directorate – see para. 6.1 of the Latham Report, HMSO, July 1994.

Apart from this, however, if the contractor considers that there are any special risks attached to the job which he cannot evaluate or are too great for him to accept, then he must make clear in his tender the basis on which he is putting forward his offer. Thus weather may be a particular hazard in the locality of the works, the stability of the rate of exchange of the currency in which payment is to be made may be doubtful, or transport to site may be totally dependent on the availability in time of certain facilities. There needs to be in the tender, preferably in or referring to the schedule of prices, a clear statement as to the assumptions on which the tender price is based, so that these can be taken into account by the purchaser at the tender appraisal stage and subsequent disputes avoided.

## Suspension

If circumstances require it, the purchaser, or the engineer on his behalf, must have the power to order the suspension of the works. The contractor should, however, have the right to claim from the purchaser for the additional costs which he is caused by the suspension (as to what these are see Chapter 22, p. 388). This right is provided for under MF/1 condition (see clause 25) but the equivalent provision in ICE conditions, clause 40, does contain rather curious limitations as to weather and the safety of the works which seem difficult to justify.

## Variations

See Chapter 22.

## Termination

There are three possible situations in which termination of the contract can occur: major default by the employer, major default by the contractor or frustration of the contract.

Since there is often considerable doubt at law as to whether or not a breach of contract is sufficiently serious to justify the injured party in determining the contract as opposed to the normal remedy in damages, it is normal to include in the contract conditions a specific right for either the contractor or the employer to terminate for certain specified breaches if these are not remedied within a stated period of notice. From the contractor's viewpoint this remedy is of the greatest importance if the employer fails to pay and it is strange that the ICE conditions do not give the contractor an express right to terminate on these grounds. The extent of the contractor's rights to terminate as a matter of law for non-payment by the employer are uncertain. Such a right depends on the breach by the employer amounting to a repudiation of the contract so entitling the contractor to decline to complete his own unfulfilled obligations. Failure to pay against one interim certificate would almost certainly not be sufficient to amount to repudiation, unless it was accompanied by other evidence which showed that the employer did not intend to make any further payments. The contractor's rights in this respect are properly covered in MF/1, clause 51.1, and in both the FIDIC conditions for electrical and mechanical works (clause 46.1) and the civil works (clause 69) although their exercise in practice is often difficult overseas where the contractor has large sums of money tied up in constructional plant in the territory concerned and staff who may be refused exit visas.

Although frustration of a contract under the English legal system is an unlikely event due to the strict requirements of English law regarding the fulfilment of contracts, and problems relating to war so far as UK contracts are concerned can be treated as largely academic, this is not the case overseas. There it is necessary to cover the situation in which performance of the contract may be rendered either more difficult or eventually impractical by reason by war, riot or similar events – as at the time of writing in Iraq. The relevant clauses in FIDIC are not unreasonable except that it is difficult to see the justification for requiring the contractor to continue with the contract when his personnel may be exposed to the risk of death or injury from enemy action. The observation by Duncan Wallace in his commentary on the FIDIC form that it is unnecessary to give the contractor a right to terminate because of the financial protection afforded to him by the remainder of the clause seems to ignore the practical realities of a construction site on which there are men, women and children exposed to such risks. It is strongly suggested that the right should be mutual.

### Language of the contract

On overseas contracts, or contracts with overseas contractors to be carried out within the UK, it is essential that the language of the contract is specified. It may be that the form of agreement including all the commercial conditions of contract are required to be in one language, usually that of the purchasing country, while the technical specifications can be in English. The language requirement needs to be established before the tender is submitted so that the costs of translation can be allowed for and arrangements made for the services of a competent and experienced local lawyer. However fluent in the local language – say Spanish – professional staff in the contractor's organization may be, they will virtually never have the necessary expertise to recognize the subtleties of the phrases being used.

## Law of the contract

It is of the utmost importance that the law of the contract should be stated in the contract so that the parties are aware of which legal system will govern their obligations and by which the contract will be interpreted. When contractors enter into main contracts with overseas clients this requirement for the express inclusion of the proper or governing law of the contract is only rarely omitted. But it can be forgotten with sub-contract where both firms are English as happened in the case of *JMJ Contractors Ltd v Marples Ridgway* [1985] 31 BLR 100. The work which involved land reclamation was to be performed in Iraq and the court decided that (1) because of the clear understanding between the parties that the sub-contract was to be back-to-back with the main contract and (2) because the contract had its most substantial connection with Iraq as the place of performance, the proper law of the contract was Iraq. This was so even although arbitration was to be English.

Since the *Marples Ridgway* case was decided the Contracts (Applicable Law) Act 1990 has been passed to give effect in English law to the EC Rome Convention. The rule has been retained that in general the parties are free to select the law to govern their contract. However, in the absence of an express choice the Act states that the contract is to be governed by the law of the country with which the contract is most closely connected. This appears at first sight to be the same test as previously applied by the English courts, but the Act then goes on to provide a series of rebuttable presumptions. The most important of these is that the law which is to be applied is that of the country in which the party who is to effect the performance characteristic of the contract has his place of business.

For an engineering or construction contract the party effecting the performance characteristic of the contract is the contractor, in the same way as under a contract for the sale of goods it is the seller. Although for contracts where the subject matter is a right in immovable property the presumption is that the contract is most closely connected with the country where the property is situate, it is clear from the commentary in the Official Journal of the European Community that this does not apply to construction contracts, since then it is the construction which is the subject of the contract and not the immovable property itself.

Would it have made any difference in the *Marples Ridgway* case if the Act had been in force at the time? The Act does allow for the presumption stated above not to apply if 'it appears from the circumstances as a whole that the contract is more closely connected with another country'. The critical factor in the court's judgement had been the linkage between the main contract, which was under Iraqi law, and the sub-contract. Given the importance attached by the court to this linkage it seems probable that the result would have been the same, but clearly the presumption under the Act would have strengthened the sub-contractor's case for English law.

With a contract involving the performance of work in an overseas territory it is unlikely that the employer, especially if a government department or other public body, will agree to any but his own law and legal system governing the contract. Broadly, systems of law outside communist territories can be divided into three categories:

1   Those which are based on the English common law even though this may to a degree have been codified. This is the type of system found in North America and much of what was once the British Empire.
2   Civil law systems based on one or other of the great codes originally issued in Western Europe during the 19th century of which the most influential has been the Code Napoléon. Such systems originally allowed only a minor role to the doctrine of judicial precedent, which forms so important a part of the English common law, although this appears to be changing, and generally place a greater emphasis on formality and the correct following of procedures than does

English law. They are found in Latin America and territories which once formed part of the French Empire or which turned to France or other continental European countries for guidance when setting up their legal systems. Such systems originally allowed only a minor role to case law but today, although still not recognizing the doctrine of judicial precedent, decisions of the courts are of increasing importance in defining the meaning to be attached to the articles of the code. Further, in the French code civil and related systems there are detailed statutory provisions dealing with the contractor's responsibilities for defects in civil and building work.

3   Those which are largely undeveloped as regards complex contractual and commercial matters but have concentrated on the law of the family, on inheritance and land.

Given the independence of the judiciary, which in certain territories would be making a large assumption, neither categories 1 nor 2 present any real problem. The law in category 2 may differ from our own but it is available and capable of definition so the employment of a skilled local lawyer should enable the contractor at least to understand the liabilities which he is assuming and the rights which he will possess. Category 3 is, however, wholly different. Not only is the law difficult to determine but its application in the case of a foreign contractor is likely to be influenced by factors of a non-legal nature. Moreover since such territories are in general governed in an autocratic manner the law can change rapidly according to the will of the ruler.

If compelled to contract under such a system the contractor must understand that he has none of the protection which would normally be afforded to him by the legal system of the UK and that he must rely on political influence rather than law in order to obtain justice.

**Taxation**
On contracts to be performed even partially overseas the contractor must be aware of the local laws relating to taxation.

Only an outline can be given here of the problems which may be encountered and expert taxation advice is necessary in each case. Points to be considered are:

1    How does the liability to local tax arise and is it possible to construct the contract(s) in such a manner that it can be avoided? Liability to tax may arise simply from having a local project office or employing local sub-contractors and if it does then the firm may find that it is taxed on the whole of the deemed profits of the contract and not just those relating to the operations performed in the country concerned. It may be possible to minimize this problem by having two separate contracts, one off-shore and one on-shore, so that local tax only applies to the on-shore work the value of which is kept to a minimum. Alternatively if the contractor already operates a local company they may be used to place local sub-contracts for local work and no project office of the parent company is established.

2    Is it better to establish for a large project a branch of the UK company or set up a local company? The advantage of the latter is that it may be possible to set up the shareholding in such a way as to take advantage of favourable taxation agreements between the local country and another country other than the UK. This used to be the position in Iran.

3    Is there a double taxation agreement between the UK and the local country and if so what are its terms?

4    Are there any local taxation incentives for participating in particular forms of activity or in particular locations?

5    Is the remittance of funds back to the UK, in particular the final retention money, subject to having received a tax clearance certificate from the local country? This can be a major source of delay in the receipt of cash in a usable form. If this is the law then the contractor must from the outset set up his operation and maintain the appropriate books of account etc. in conformity with local practice and employ as his auditor a firm acceptable to the tax office.

6    Is tax payable on actual profits earned or on deemed profits and if so how are the latter calculated? Can charges made by

the UK company to its local offspring, say for technical services, be set against local tax? Many countries have now become wise to this and any allowance for such charges is often minimal.

7   Is it possible to obtain tax exemption for the contract on the basis that it is being carried out for the government? The contractor needs to be very careful about accepting any promises as to tax exemption. Often the ministry or authority concerned has no right to grant such exemption although it will be tempted to say that it has in order to obtain a lower price. The contractor will then later discover that the finance ministry will claim the tax they consider due and refuse to take note of what is written in the contract. Nor is it wise to rely on an indemnity from the purchasing ministry or authority against the tax due since when the time comes to pay they are unlikely to have the funds.

There are only two safe ways of dealing with this problem. One is to obtain a tax exemption certificate from the finance ministry. The other is to include an allowance for the tax in the contract price as a provisional sum with an undertaking to repay any balance.

# 13 Contract price

There are broadly three ways in which the contract price may be expressed or calculated:

- lump sum
- schedule of rates or bill of approximate quantities
- cost reimbursement.

These different ways are not necessarily mutually exclusive. Thus the above-ground element of a building contract may be on a lump sum basis whilst the foundations are subject to remeasurement; the supply portion of a plant contract may be a lump sum, whilst the installation of the plant is on cost reimbursement; a contract for a complex chemical plant may be on cost reimbursement but with the overheads and profit margin compounded as a lump sum.

The choice of which way to ask the contractor to price the work will depend very largely on the amount of information regarding the job, and the conditions under which it will be carried out, which the buyer can provide to the contractor in the time available for tendering.

# Lump sum

The nature of a lump sum contract has come before the courts a number of times on the issue as to whether or not entire performance of the contract was a condition precedent to payment. In general, the courts have leaned towards the construction that, provided the contract has been substantially performed, even if imperfectly, then the contractor will be entitled to payment of the contract price less an amount for the remedying of defects (*Hoenig v Isaacs* in the Court of Appeal [1952] 2 All ER 176).

From the purchaser's point of view the firm lump sum is ideal. It establishes the amount of his commitment in advance, it provides the maximum incentive to the contractor to complete the work on time, and it reduces to a minimum the amount of administrative work involved after the contract has been let.

Under a pure lump sum contract the contractor will not be entitled to any additional payment if work indispensably necessary to complete the contract is omitted from the specification. Further in cases largely decided in the last century it was held on a number of occasions that under such a contract the contractor was not entitled to additional payment if additional expense was incurred in order to fulfil the contract because of errors in the plans, specifications or information provided by the employer at time of tender. However the validity of such decisions today must be doubtful in the light of cases such as *Hedley Byrne v Heller Partners* [1963] 2 All ER and others on the liability of employers for negligence in the provision of information and the Misrepresentation Act 1967 as discussed in Chapter 3.

The benefits referred to above will accordingly only be obtained if it has been possible for the employer to provide the tenderers with appropriate and accurate information on which to base their tenders and as regards other information which the tenderers must obtain for themselves if it has been possible for them to obtain it in the time available.

Further it is essential in a lump sum form of contract for the employer already to have made up his mind what he wants and for subsequent variations to be minimal since the contract itself will in general provide no mechanism for the pricing of variations – see further Chapter 22.

A checklist of the general questions to which a tenderer required answers when bidding has already been given (see pp. 108–9). In order to be able to tender on a lump sum basis the estimator must have answers to the following either from the employer or his own company:

1   Material quantities and specifications. These may be in the form of drawings from which the estimator can himself take off quantities.
2   Tolerances permitted and any special finishes required.
3   Labour hours and trades both for shop production and on site. This means that decisions on methods of production/construction affecting labour quantities and skills must have been made.
4   Description and quantities of bought-out items. This requires decisions to have been taken on, for example, sizes, capacities, and horsepowers.
5   Types of production or constructional plant which will be utilized both in the shops and on site, and the times or periods involved.
6   Where design is significant, and is not included as an overhead, the amount of design work involved.
7   The site organization which will be needed and for what period.
8   Overtime to be worked in shops and on site.
9   Time when the work is to be carried out.
10  Factors which will affect labour productivity on site – climatic conditions, religious holidays, nationality of labour to be employed.
11  Geographical and climatic factors as they affect civil, building or mechanical and electrical site work. These would include rainfall, presence of corrosive salts liable to attack steelwork, humidity, dust, availability of fresh water, general local facilities, supply of clean aggregates.
12  Local material availability, e.g. cement in proper condition

and in the right quantities to meet programme, port offloading and transport facilities including any heavy load restrictions on roads or bridges.

13 General local amenities and workshop facilities.

This is a formidable list. It confirms the need for the purchaser to be able to give complete and accurate information before a firm lump sum price can be tendered. It also indicates the time and cost in which the contractor is involved in lump sum tendering. What must be remembered is that every time a tenderer guesses, he may guess wrong, and every wrong guess costs someone money. Moreover that someone, if the tenderer is to stay in business, can in the long run only be the employer whether on that particular contract or another.

Just as the contractor's problem on lump sum tendering is to assess the risks involved, so the employer's problem is the time which it will take him to give the information necessary to reduce those risks to reasonable proportions. Some element of risk there will always be; that is in the very nature of contracting itself.

The problem of information against time arises particularly on contracts for building and civil engineering work where the employer is normally, though not necessarily, responsible for design, and two of the main factors affecting design are both largely outside the designer's control. These are, first, the nature of the subsoil and, second, the detailed requirements of the specialist contractors and sub-contractors for plant and services. Increasingly, these latter form a major part of most building or civil engineering projects. If the start of construction were to be delayed until exhaustive bore-hole research had been carried out and detailed designs for the plant and services prepared, the element of uncertainty could of course be minimized. Managements, however, are not normally prepared to accept delays of this sort so that it becomes necessary to find some way in which the risks inherent in these unknowns can properly be shared between contractor and employer and a start made on the project.

# Schedule of rates or bill of approximate quantities

This leads to the schedule of rates or bills of approximate quantity method of pricing under which a schedule or bill is prepared, covering each of the items which it is anticipated may be met during the course of construction, for example excavation, concreting, brickwork, etc. These items are then priced by the contractor and he is paid at those rates for the amount of work actually carried out, irrespective of the quantity shown against the item in the schedule or bill. The problem has however always existed of where the change in the quantity from the estimated to the actual is such that it affects the contractor's method of working, e.g. a change to hand from machine work. It was expressly provided in the 5th edition of the ICE conditions and the provision retained in the 6th edition that the contractor is entitled to an adjustment of the rate if there is a change in quantities which make the rate 'unreasonable or inapplicable', and there is no minimum percentage change required. In *Mitsui Construction Co. Ltd v Attorney General of Hong Kong* 10 CON LR, it was decided by the Privy Council, on wording similar to that of clause 56(2), that it was immaterial whether the change in quantities arose from a variation order or not. All that was necessary to give the engineer a discretion to agree or fix new rates was that there was a sufficient discrepancy between the billed and measured quantities which on the facts of that dispute was demonstrably the case. It also held that the sufficiency of tender clause which was in similar wording to the ICE clause, did not put the risk of adverse ground conditions on the contractor, so as to prevent the engineer from exercising his powers under the equivalent of ICE clause 56(2).

In pricing a contract in this way a contractor has to estimate the quantity and cost of the labour, materials, and plant which will be required to execute the given quantity of work. Since the major elements are labour and plant, the assessment of productivity is a vital part of the estimating process. This in turn is closely related to the physical conditions under which the work will be carried out – for example, the time of year – and to the possibility of carrying out the work in a planned way with a reasonable degree of continuity – for example, drawings arriving on site well in advance of the commencement of construction of the work to which they relate. The importance of these points will be referred to again when discussing variations and claims.

As regards specialist sub-contractors' work, these items are made the subject of prime cost or provisional sums. An amount is included by the employer in the bill which represents his best estimate of the cost of the item. When the sub-contract is placed (after the main contract has been let) that sum is deleted and replaced by the amount of the sub-contract. When tendering himself, the main contractor is only required to tender the margin he wants for handling the sub-contractor, usually expressed as a percentage plus any sum he wants for attendance on the sub-contractor, like providing scaffolding, storage, etc.

# Cost reimbursement

With many industrial projects today, speed in getting work carried out is regarded as more vital than lowest initial capital cost. Moreover, apparent cost advantages at tendering stage may be lost by the time final settlement is reached on the payment of claims. On the other hand, simple cost reimbursement provides no incentive to the contractor to minimize costs, nor any penalty should he fail. Indeed the reverse is true. Most contractors in fact dislike straight cost plus because of the inefficiencies which it may breed within their own company. Costs can so easily be charged to cost-plus jobs if no other home can be found for them!

Various types of incentive, target cost or co-operative forms of contract have been devised, therefore, as a means of combining the flexibility and speed associated with cost reimbursement with a strong measure of costs discipline and an incentive to efficiency and economy.

All these forms of contract have certain features in common:

1 The principle of design and construction in parallel as opposed to in series.

2 The early establishment of a target estimate either as a definite sum or on civil or building work as rates in an approximate bill of quantities, against which the work can be remeasured.

3 The recording of the actual costs incurred and their comparison with the final target cost. This is the original target cost adjusted to take account of authorized variations.

4 The sharing between employer and contractor of the difference between 2 and 3.

5 The payment of a lump sum in addition to costs which will cover at least the contractor's head office overheads and profit. Additionally there could be included other items such as design charges, procurement charges and even site project management if there was sufficient information available for these to be estimated on a lump sum basis.

How the final contract price is arrived at under the conventional, and the target or incentive form of contract, can best be illustrated by Figure 13.1.

Two points need particular attention at the negotiating stage. First, the division between the employer and contractor of the difference between the target cost and the actual cost which may be either a saving or an over-run. In its simplest form there is a single percentage split for the over-run and another for the savings. Alternatively the percentages can be adjusted on a scale according to the amounts above or below target. Preferably with this latter scheme the contractor should be required to accept 100

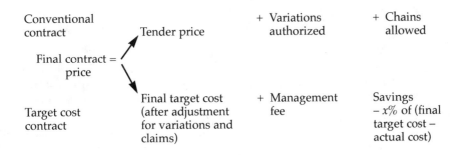

Conventional contract — Tender price — + Variations authorized — + Chains allowed

Final contract = price

Target cost contract — Final target cost (after adjustment for variations and claims) — + Management fee — Savings $- x\%$ of (final target cost − actual cost)

**Figure 13.1  How the final contract price is arrived at**

per cent of the cost over-run above a certain limit, but this will only be feasible if the contractor can make a reasonable assessment of the risks involved. Cost and time for completion can also be integrated as described on p. 266. This ceiling may be the target itself or more likely the target plus a certain margin, the extent of which will reflect the unknowns inherent in the contract. Second, in the assessment of the target cost it is essential that the target should be built up from the component elements of labour materials, plant, etc., which the contractor can be expected to use on the job, and has regard to the construction or manufacturing methods which it is anticipated that the contractor will adopt. It is not just a question of selecting 'average' competitive rates, but of seeing that they are tailored to the job in question and reflect its particular circumstances. The target must, however, contain a contingency margin which is sufficient to ensure that, provided the contractor uses proper efficiency, the target remains at all times credible to beat. The aim should be to set a target which ought to be beaten by a low margin, say 10 per cent.

Target cost contracts are notoriously difficult and expensive to manage and administer. Variations are bound to occur and so are the arguments as to whether something is a change or not. Each variation will mean negotiating a change to the target and possibly the management fee. Costs must be recorded and audited and deductions made for re-work which is due to the contractor's default. With the emphasis placed on speed it is only too easy for the administration and control to suffer so that the commercial side lags far behind the execution of the works. If this happens the whole benefit of the target cost mechanism as providing an incentive will be lost and claims will be inevitable.

In certain instances it may be preferable, rather than using a target cost form of contract, to negotiate a basis for cost reimbursement with the intention that this should apply until the point is reached in the definition of the project at which it is possible to negotiate a lump sum price. A suggested way in which the various elements of the contract price should be dealt with in establishing the basis on which costs are to be recovered is set out below. The comments made may also be appropriate to the negotiation of target cost contracts.

## Design

This is usually paid for on a man–time basis, the unit of time – hour, day, week, month, or even year – being selected to suit the individual contract. Rather than be concerned with the actual salary of the individual draughtsman or engineer, it is often convenient to establish an average salary for a particular grade. The following points need watching when considering these rates:

1   In respect of which classes of staff are they payable? This may be only actual engineers or draughtsmen or may extend through bills of material clerks to clerks, typists, and the like. Obviously this alters substantially the allowance for overheads; the smaller the chargeable base the higher the overhead.
2   Are the overheads included in the rates, the whole of the company's overheads, or only those related to design? Practice differs on this according to whether the firm's normal selling unit is design time or not. If it is, then normally all overheads (other than possibly those relating purely to construction or procurement) will be charged against design.
3   The above two points have a tremendous effect on the overhead as a percentage. The swing can be as much as from 75 to 300 per cent.
4   Do the rates include:
     ●   overtime
     ●   travelling and subsistence
     ●   telex, cables and telephone calls
     ●   printing and reproduction costs
     or are these chargeable at net cost?

5   Do the same rates apply to sub-contract design?

Obviously from the employer's point of view the more elements which can be properly made the subject of lump sums the better;

particularly if the job is going out to competition. It is extremely difficult to compare either percentages or hourly rates; percentages because these have no validity by themselves but only when related to a base, and it cannot be assumed that the base will be the same for all firms; hourly rates because these have no validity unless one is in a position to assess the real value to be placed on the work which will be turned out in an hour, and quite simply one is not.

Thus firm *A* may offer design at £27.50 an hour, firm *B* at £29.50 an hour. But by themselves these figures mean nothing. Firm *A* may take 50 000 man-hours and produce a design which costs £1 250 000 to build. Firm *B* may take only 35 000 man-hours and their scheme results in a final price £1 150 000. The same sort of reasoning applies to labour rates for construction or erection work.

### Procurement

This is usually paid for as a percentage of the value of materials purchased after deduction of trade but not cash discounts. It includes purchasing, expediting and inspection. Again one needs to check that travel and subsistence, which may be high, are included.

### Materials

Net price after deduction of trade but not cash discounts. The total value of discounts can be very substantial, particularly on items such as motors, valves, pipework, and so on and should not be regarded as the estimator's contingency.

### Site supervision, UK contracts

This may be negotiated as a lump sum, or a weekly rate. It will include:

1   Salaries and allowances for staff which will include:

salary
site allowance
national insurance

pension contribution
company car (where this is provided)
medical insurance
employer's liability insurance
CITB levy (where applicable)
expenses (for senior staff)
periodic fares.

Again it is normal, rather than dealing in individual salaries, for rates to be established for various categories of staff expected to be employed.

2   Offices and stores either on a rental or build basis.
3   Office running costs often including provision for computers.
4   General site transport.
5   Consumables.
6   Canteen.

## Erection labour, UK contracts
Charges for erection labour on a per hour basis will normally include:

1   Wages and allowance – for example, subsistence and radius allowance, condition money, etc.
2   Bonus.
3   National insurance, holiday with pay, redundancy fund payment, etc.
4   Common law insurance.
5   Hand tools.

Care needs to be taken in dealing with the non-productive element of overtime. This will affect only a small proportion of the overhead charges related to wages.

## Site supervision overseas
On overseas contracts the indirect charges for supervisory staff may easily amount to 150–200 per cent of the man's payroll costs. Such charges may include:

1 Provision of accommodation, its maintenance and services costs, e.g. charges for electricity and water which are often substantially higher than in the UK.
2 A car and its running/maintenance costs. Although, depending on the territory, petrol may be cheaper than in the UK, maintenance and depreciation may be very high due to unfavourable climatic conditions and the poor state of the roads.
3 Food allowance.
4 Air fares to and from the UK.

Practice on housing and food varies both from company to company and according to the size of the supervisory team.

On projects only involving a small team some firms pay a fixed allowance per day leaving it to the man to find his own accommodation and food. This is often preferred by the supervisory staff since it leaves them free to choose their own standard of living and sometimes to form local 'liaisons'. It does however weaken the company's control and may lead to staff living and behaving in a manner which lowers the firm's reputation.

With larger contracts it would be normal for the company either to set up a camp containing shopping, laundering and recreational facilities in addition to the accommodation or select itself the accommodation for its staff according to their grades. In remote areas or where there is no suitable expatriate type of accommodation, there is no alternative and the company must totally establish its own facilities, often including drinking water, sewerage, catering facilities, cottage hospital, etc.

A further complication is caused on long-term contracts by the need to offer at least senior staff married contracts. There are then the problems of children's education, either paying fees for a local English/American school if there is one or if not providing a school with a teacher for juniors and paying at least part of boarding school fees for older children. Whilst bachelors can be accommodated in flats or barrack-type blocks on site, families need houses. If suitable ones do not exist locally then pre-fabricated ones, air-conditioned and complete with all services, must be provided. Shopping and medical and recreational facilities will need to be expanded to cater for the needs of wives and children or additional allowances paid to allow use to be made by families of local facilities which are normally expensive. Air fares to the UK for mid-term leave and for children at boarding school for at least one holiday per year with parents need to be included.

Finally if the overseas territory imposes income tax there may be the vexed question of taxation to be considered, and the following issues will arise:

1   Is there a double taxation convention between the overseas territory and the UK? If so what are its terms?
2   On what basis will the individual be charged tax in the overseas territory – on his living allowance only or his total earnings as assessed by the local tax authorities?
3   Does an exit visa from the territory depend on the issue of a tax clearance certificate?

Although firms often take the line that while giving general answers an individual's tax affairs are his own business, they inevitably get drawn into the problem since the individual who goes to work overseas is only interested in the nett remuneration package. American companies since the introduction of legislation taxing a person in the USA on his combined American and overseas earnings including benefits have found this out to their cost.

To the extent that a contractor is able to utilize locally recruited administrative and professional staff then the costs of supervision will necessarily be reduced by the allowances for housing, accommodation, air fares etc., although in territories in which such staff of an appropriate calibre are available base salaries are not likely to differ widely from those in the UK. Depending however on the nationality of such staff, problems may however arise in terms of differences in social and religious habits and customs and these will be accentuated if both the ex-patriate and local staff are required, because of the site's isolation, to live as well as work together.

## Overseas erection and construction labour

On overseas contracts construction and erection labour will be either local or recruited from third national countries, e.g. in the Gulf States and Saudi Arabia labour including skilled tradesmen will come from either India, Pakistan or the Philippines. Different considerations, all of which affect pricing, apply to the employment of local as opposed to third national labour and may briefly be summarized as follows:

### Third national labour

1 Recruited through a labour contractor in the country of origin on a one- or two-year contract.
2 Trade testing normally carried out in country of origin.
3 Payment includes recruitment fee and return air fare.
4 Wages are generally subject to control of the government of origin.
5 Accommodation and food must be provided.
6 Work permits and visas are usually restricted to employment by the contractor and the man must return to his country of origin on termination of his contract.

In practice at the time of writing the employment of third nationals on the above basis has had the effect of largely protecting the contractor against escalation or labour disturbances but has also meant that he lacks flexibility in being able to hire and fire. Labour costs become in effect semi-fixed instead of a variable which could have a major impact if the work programme becomes subject to changes or delays outwith the contractor's control. Productivity becomes largely a function of the number of men on the site relative to the available quantity of work.

### Local labour

In general it is not recommended that contractors working overseas should directly employ local construction/erection labour for the following reasons:

1 Local labour laws which can be tough in their theoretical

provisions will be strictly enforced against an ex-patriate contractor and complying with them to the letter in terms of working hours, redundancy payments bonuses, etc. will be expensive. Local contractors however have a way of getting round these provisions or at least minimizing their cost impact.

2 Local working and amenity practices will be unfamiliar to the expatriate contractor and even when he does become aware of them they will be difficult for him to apply, which again will cost the contractor money.

3 Trade testing and qualifications may not exist.

4 The problem of language.

5 Local contractors will already have an established network of relationships with government and union officials, client's inspectors etc. and also with sources of reliable labour which an expatriate firm on its own will never achieve.

The only circumstance under which an expatriate firm can successfully employ any quantity of local labour is if it has formed a joint venture with an established local contractor and matters related to the employment of local labour and dealing with local officialdom are made the responsibility of the local partner who is obliged therefore to take an active interest in the partnership.

## Constructional plant

There will normally be a schedule of weekly hire rates. The following points need covering:

1 Do the rates include any element of profit?

2 Are they tied to a number of hours?

3 Do they include charges for driver?

4 Do they include fuel, lubricants, spares, maintenance? There is a danger of paying twice.

5 Do they include charges for transport to and from site? These are often too heavy.

6 Where the plant belongs to the contractor, what allowance has been made for depreciation and what residual value has

been assumed?

On overseas projects, unless exemption is granted by the government, the problem may arise regarding import duties. First, even though the plant is only being imported temporarily duty may be payable on certain types of plant or on particular makes in order to protect local industry or exclusive dealer arrangements. Even if this is not the case, duty or a bond in lieu may require to be deposited which is in theory returnable on that item of plant being re-exported but is forfeit if the plant is sold in country. Unfortunately in practice temporary importation procedures tend to be so drawn out that in desperation to get moving the contractor will pay the import duty and hope to recover it. The actual practice of temporary importation in the territory in question needs to be carefully examined both by the contractor and by the employer's negotiator.

## Management overheads and profit

Preferably a lump sum which can be made the subject of competitive tender. Sometimes, depending on the information available, it may be possible to include in this lump sum the design element and even perhaps the site supervision, leaving only the direct materials, sub-contract, and labour costs to be either reimbursable or negotiated during the contract period.

## Price to be agreed

Each of the methods of pricing referred to above has the merit that, although the final price may not be established when the contract is let, at least the mechanics for doing so and the principles to be followed have been settled. There are, however, occasions on which the contracts officer finds himself urged, in order that work can get started, to place the order or contract simply on the basis of 'price to be agreed'.

The problem with this type of arrangement is that there cannot be a 'contract to agree' anymore than there can be a 'contract to negotiate' (*Courtney & Fairbairn Ltd v Talaini Brothers (Hotels) Ltd* [1975] 2 BLR 97, now confirmed by the decision of the House of Lords in *Walford and Others v Miles and Another, The Times* 27 January 1992). Price in a constructional contract is so essential that in order for a contract to exist there must either be a price agreed or there must be an agreed method of ascertaining it otherwise than by negotiation between the parties. In order to overcome this difficulty it is suggested that a letter of intent along the lines of that suggested on p. 157 should be issued together with programme for the negotiations for the completion of the contract.

As an alternative, if the parties are willing to agree then following a suggestion made by Lord Denning when giving judgement in the above case, a third party, say an independent quantity surveyor, could be given authority within a fixed period when it was considered that sufficient data would become available, to settle prices which he considered were fair and reasonable, perhaps with specific instruction on the level to be allowed for profit.

# 14 Terms of payment

## Policy considerations

Terms of payment are a matter on which the commercial/technical and financial sides of the employer's business may find themselves pulling in opposite directions. The employer may attain the best commercial and technical result if he offers to the tenderers terms of payment which, while providing the employer with reasonable contractual safeguards, impose the minimum strain on the contractor's financial resources. By so doing the employer will:

1  Avoid having to restrict the tender list to large firms possessing the resources to finance the contract, whose overheads and prices will be higher than those of smaller companies. (This assumes of course that such smaller companies are otherwise technically and commercially competent to carry out the work.)
2  Ensure that the tenderers do not have to inflate their tender prices by financing charges. In many instances the rate of interest which the contractor has to pay when borrowing will

be higher than that paid by the employer.
3   Give encouragement to, and be able to take advantage of, firms possessing technical initiative who would otherwise be held back from expanding by lack of liquid cash.
4   Minimize the risk of being saddled with a contractor who has insufficient cash with which to carry out the contract and of having, therefore, either to support the contractor financially or terminate the contract.

On the other hand, to offer such terms means that the employer has to finance the work in progress and tie up his own capital in advance of obtaining any return on his investment. Particularly with a project such as a new factory or power plant, it would impose the least strain on the employer's financial resources if he could avoid having to pay anything at all until the project is earning money, and make the payments wholly out of revenue. With very large contracts of this type overseas, particularly in the underdeveloped countries, buying on credit in this way is not a matter of choice but of necessity. The authorities or companies concerned are not in a position to do anything else. As usual, however, the price which a customer pays for credit is high. Even with preferentially low interest rates for exports the cost to the purchaser of the financing charges on a long-term credit contract may easily amount to a third of the 'cash' selling price.

The factors related to cash flow and contract risk/profitability on both home and international contracts were analysed by Roland B. Neo in *International Construction Contracting* (Gower, 1976). However, only construction contracts under which payment is related broadly to the value of work carried out were considered and on overseas contracts the effect of having to provide bonds cashable by the purchaser on demand was not taken into account, a factor which seriously diminishes the author's conclusions on risk assessment.

On civil engineering and building contracts carried out in the UK either under the ICE or JCT forms or some major customer's adaptation of these, the contractor is paid monthly for the value of work done and materials delivered to site for incorporation into the permanent works in the preceding month, less a percentage for retention money. The relationship of the main contractor's cash expended to payment received will be determined largely by the relationship between the work which is carried out by the main contractor, that which is undertaken by domestic sub-contractors and that which is performed by nominated sub-contractors. Since the last edition of this book there has been an important change in the way in which contracts are performed. Today in most larger contracts there is little if any work actually performed by the main contractor utilizing his own labour. The functions performed by the main contractor are limited to design of the temporary works, provision of perhaps certain site facilities and the planning, co-ordination, management, supervision and administration of the contract, with the work being carried out by domestic sub-contractors and to a lesser extent by nominated sub-contractors. Even constructional plant will normally be hired.

As a result the main contractor is in a position markedly to improve his cash flow by delaying payments to his sub-contractors. In times of recession and intense competition with low, if any, profit margins built into the tender price, conditions which at the time of writing have prevailed for some time in the industry, the main contractor has often had to depend on interest earned from delayed payments to sub-contractors, together with claims, for making a profit. For many firms the temptation to delay payments in this way has been irresistible.

In addition it has become normal for main contractors to insist on 'pay when paid' clauses so that, if there is any delay by the employer in payment, or any deductions made by him, a sub-contractor, even if not himself in default, is likely to suffer.

The curve shown in Figure 14.1 of the previous edition is therefore no longer appropriate. Rather than showing a cost to the main contractor of financing a contract, such a curve today would be more likely to show that the main contractor has actually earned a profit.

A possible scenario under which this would happen is shown in the updated Figure 14.1 which is based on the following presumptions:

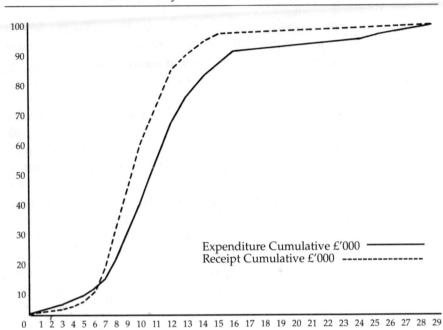

**Figure 14.1   Contractor's expenditure in relation to payments received**

1   Contract costs total £1 million, contract period 12 months.
2   20 per cent of the contract costs relate to the contractor's own work.
3   60 per cent of the contract costs are represented by work performed by domestic sub-contractors.
4   20 per cent of the contract costs are represented by nominated sub-contractors or suppliers.
5   Payment is received by the main contractor within two months after the end of the month in which the work was performed.
6   The main contractor pays his domestic sub-contractors within two months after he himself receives payment. For work therefore executed by the sub-contractor in month 2 it has been assumed that payment will be made by the main contractor during month 5.
7   The main contractor pays his nominated sub-contractors

and suppliers within one month after he himself receives payment.

8    The main contractor holds back retention money of 5 per cent of the sub-contractor's prices until the same periods after he himself receives payment from the employer.

On the above basis it is estimated that the main contractor would increase his profits by approximately £8000 assuming that he can earn 8 per cent interest on the money withheld.

Roland Neo mentions but does not take account of two other actions which a main contractor can take to improve cash flow: over-measurement in the early months of the contract and front-end loading by artificially increasing the value of the rates for the work to be carried out early. These practices are common both in the UK and more especially overseas and unless carried to excess the engineer or quantity surveyor will often turn a blind eye to them as being a matter of custom and practice. Indeed during a time of high or sharply rising inflation, if the contract is subject to contract price adjustment they can work to the contractor's disadvantage by diminishing the amount of escalation recovered relative to that incurred (see further Chapter 20).

The other factor which may materially affect cash flow is the extent of variation orders issued by the architect/engineer which are not covered by rates and prices contained within the bills of quantity and the time taken to get such rates and prices agreed. Although payment for such variations will normally be made on interim certificates on a provisional basis the amount certified will inevitably be conservative.

It has been assumed in the example in Figure 14.1 that payments are made by the employer in due time, the final account is settled promptly and that there are no important issues which result in the right of set-off being exercised by the person due to pay against the party entitled to receive payment. Unfortunately only too often these assumptions would not be true and the Lathan Report was fully justified in its criticisms of the practices on these issues prevailing in the industry.

On overseas construction contracts for which the contractor

can expect to receive a down payment of say 10 per cent of the contract price it might be expected that the amount of 'capital lock-up', to use Neo's term, would be less, indeed that is his suggestion. Whether however this is the case in practice will depend on the following factors, several of which Neo does not take into account:

1   The amount of the contractor's initial expense for such items as ECGD premiums, bonding charges and agents' fees which are payable on contract signature. These may easily amount to 10 per cent of the contract price.
2   Mobilization costs in the overseas territory, e.g. setting up the site establishment, importation of plant, and whether these are paid as a separate item from the down payment, are deemed to be covered by the down payment or amortized over the billed rates and recovered pro-rata to progress. Only if they are paid as a separate item is the contractor's cash flow likely to be other than negative particularly on any contract such as road construction which is plant-intensive.
3   Delays in certification and payment either deliberate or the result of bureaucratic inefficiency. This applies especially to payments for escalation or variations. Certification delays may be mitigated if the contract is being supervised by international consultants but those for payment can usually only be reduced by personal attention being given to each person in the chain of required signatures to ensure that your piece of paper is moved ahead of others.
4   Delays in the release of retention monies usually due to the unwillingness of overseas clients to take the responsibility of releasing their hold over the contractor.

Although not directly affecting the cash flow situation of the individual contract the contractor's financial position as a whole will be materially influenced by any on-demand bonds which he is required to put up under the terms of the overseas contract. The value of these will be regarded by his bank as liabilities when deciding on the extent of the facilities they are prepared to make available to him. A contract of say £20 million on which the amount of bonding averaged 15 per cent could therefore reduce the contractor's overdraft limit of £3 million.

Considerable dissatisfaction has been expressed recently in the construction industry, especially by employers, with the traditional methods of monthly payments. This is for the following reasons:

1 It offers little incentive to the contractor to progress the works or meet interim dates which are of critical importance to the employer.
2 It largely transfers the burden of financing the work from the contractor to the employer and so allows for firms to establish themselves as contractors with very little in the way of capital and therefore unable to fund expenses such as training or to meet their commitments for defective work.
3 The main contractor has an incentive to retain the interim payments made to him in respect of the work of sub-contractors for as long as possible in order to improve his profit margin. Because material suppliers usually require payment within limited credit terms and specialist firms at least are in a position to enforce these, this forces specialist sub-contractors to finance their work.
4 It is time-consuming and expensive and a source of much conflict between the parties because of the subjectivity of the assessments made as to the percentage complete of the items of work involved.
5 If the main contractor goes into liquidation during the course of the contract, the employer having effectively paid out money in advance against completion is unlikely to be able to recover the additional costs he incurs in having the work completed, unless he has the security of an adequate on-demand performance bond.

The Latham Report goes so far as to recommend the phasing out of the system of monthly valuations (recommendation 8, para. 5.18).

The only practical problem related to this is the need for a very clear definition of the events against which payment is to be made in order to avoid time-consuming disputes which will defeat the whole purpose of the exercise. For plant and equipment contracts it has been traditional within the UK for payments only to be made against the value of materials delivered to site so that the burden of the financing costs during manufacture has fallen either on the contractor or the actual manufacturers of main items of plant and equipment. Continental practice has however for a long time been to make payments in stages during manufacture and this practice has to a degree spread within the UK. However, more recently certain employers have reversed this practice even to the extent of proposing that there should be no progress payments at all but only one payment on completion, except for the retention money. The argument advanced for this is that it imposes a strong incentive on the contractor to complete on time.

In the author's view objection can reasonably be taken to the principle of expecting the contractor wholly to finance the work either up to delivery or to the commencement of site erection. First, it imposes on many companies a substantial and continuing strain on their cash flow position. This is particularly so when the company is seeking to expand or to take contracts of longer duration. Second, in most circumstances it is cheaper for the employer to borrow money than for the contractor to do so. Third, the ownership of work constructed on the employer's site belongs to the employer and cannot be used by the contractor as security against which he can borrow money. This is in contrast to a manufacturer where work in progress in his factory belongs to him. With private sector clients therefore the contractor would be certain to require security for payment which was in priority to the employer's normal business creditors. This would again cost the employer money. Fourth, it is against the national interest in that it puts such firms at a disadvantage when competing against continental companies who would normally expect in their domestic market to be paid as much as 30 per cent of the contract price with order. Clearly such companies do not have to include within their overheads for the financing of work in progress; their money is turned over faster, capital employed is reduced, and they can invest more, for example, in development and new machinery. These are formidable advantages.

So far as retentions are concerned, the purchaser is obviously sensible to withhold an appropriate percentage of the purchase price until he is satisfied that the plant is working properly and has met its guarantees. Provided that the retention moneys are considered in this way and not as a form of finance for the purchaser then the contractor can have no reasonable objection to them.

To sum up, provided the employer can possibly afford to do so he is likely to get the best bargain if he, rather than the contractor, largely finances the contract. By so doing he will also be acting in the national interest and indirectly therefore in his own, by assisting British companies to compete abroad on level terms.

# Contractual safeguards

In order to safeguard the interests of both parties the contract should:

1  Define precisely the events against which payment becomes due.
2  Relate those events to the achievement of some particular objective.
3  State the amount due at each stage or provide a mechanism by which such amount can be determined.
4  Establish a time limit within which payment must be made.
5  Provide the contractor with an effective remedy should the employer default in payment.
6  Provide the employer with means by which he can obtain or recover the value of payments made before completion should the contractor default and be unable to complete.

### Definition of events (1 and 2)
Where the contract includes for the issue by the nominated engineer of certificates, then provided the criteria for these have been properly established no problem should arise unless for any reason, other than the contractor's default, the issue of a certificate is delayed. To cover this possibility two provisions are required:

1  The certificate must be issued within a stated time of an application which the contractor was entitled to make.
2  If issue of the certificate is delayed because the event itself is delayed, that is guarantee tests cannot be held because employer's other work is not ready, then after a suitable time the contractor must become entitled to the payment. The same applies in relation to delayed delivery because of non-readiness of the employer to receive the goods.

If, however, entitlement to payment is to be determined solely by reference to an event, for example delivery of the goods f.o.b., together with relevant shipping documents, then it is important if misunderstandings are to be avoided to ensure that the event is clearly described and that it is kept simple.

It is desirable to avoid multiple requirements wherever possible, since it will often be found in practice that one of them takes much longer to comply with than the others.

From the employer's point of view it is inadvisable for payment to be related solely to time, for example six months after placing the order, unless this is qualified by a requirement as to the progress which must also have been achieved. This can be done very simply by providing that the engineer must be satisfied that progress is to programme or, if not, then he can reduce the amount to be paid.

Alternatively, if the work is being controlled through network analysis, values can be allocated to certain key activities, and the contract can provide that payment of these sums will be made as those activities are completed. Properly planned, the linking together of payment and programme in this way can be most effective.

A problem which can arise on the sums due on commercial operation or take over is that often the contractor has carried out all but a small amount of the work involved but, because there is still some work outstanding, the engineer is unwilling to issue his certificate, so that retention money to the value of very many times the outstanding work continues to be withheld. Provided what has still to be done does not significantly affect the operation of the works, there is no reason why the engineer should not issue the certificate with an appropriate endorsement and release the retention money, apart from whatever he considers appropriate to retain in order to ensure satisfactory completion of the outstanding work. This is specifically provided for in 40.1(3) Terms of Payment of MF/1).

## Determination of amount due (3)

Only rarely will the contract state a definite sum to be paid at the various stages of completion; usually it will refer only to percentages, for example:

10% with order
80% on delivery
5% on take over
5% on final acceptance.

As with any percentage, it is important that no ambiguity should arise as to the base to which it relates. On supply and erection contracts there are broadly two possibilities:

1  All percentages relate to the contract price as a whole.
2  The percentages due on delivery are calculated on the contract value of the materials delivered (excluding therefore the erection and commissioning element of the price), and those elements are paid for separately as the work is carried out. In that event the 80 per cent payment might be expressed in the contract as 80 per cent of the value of materials delivered to and work executed on site (see, for example, condition 40.1(b) Terms of Payment of MF/1).

The contract should also clearly establish the method of payment for variations and price escalation.

## Variations

It is suggested that variations should be paid without any down payment and that the down payment is recovered therefore only against the original value of the contract. Retention money however would normally be deducted from the value of the variations executed.

### Escalation

If the contract is subject to contract price adjustment then it is essential to establish the data necessary for the calculation of the amount of escalation due on the variation unless for simplicity the price for the variation can be settled on a fixed price basis. Payment for escalation, it is proposed, should be made with each monthly certificate at 100 per cent of the value properly claimed. There seems no justification for involving escalation payments with either the recovery of any down payment or percentage deductions for retention.

Care however needs to be taken in the contract drafting particularly in respect of the use of the term 'contract price'. If the contract price is defined as 'the sum named in the contract subject to such additions thereto or deductions therefrom as may be made under any provisions of the contract' and the term contract price is then used in the payments clause without qualification it could be argued that both down payment and retention provisions apply to variations and escalation alike. It is preferable to set out separately the payment terms for both these items so that no ambiguity can arise. In fact the ICE and JCT conditions retention is withheld from payments made for escalation and the argument for doing this in relation to the ICE conditions is that the contractual entitlement to any payment for escalation is derived from the payments clause 60 (2) and the amounts certified by the engineer under this clause are subject to retention.

### Time limit for payment (4 and 5)

No one likes paying bills before they are obliged to do so. The accountants for big companies have been quick to see the money which can be saved by not paying their creditors until the last day for payment (unless a discount for cash has been offered). The short-term investment of daily cash balances can make a useful contribution to company profits. The administrative procedures of large organizations, both public and private, can of themselves impose substantial delays in the money actually being paid. Main contractors, to protect their own position, have developed the habit of only paying sub-contractors on 'as and when' terms, that is when they themselves have been paid by the employer.

All this emphasizes the need for the contract conditions to lay down a clear time limit within which payment should be made, which is practical in the circumstances of the contract. It is better to lay down a rather longer time initially, which stands a reasonable chance of being kept, than to include the standard 28-day clause knowing that it is unlikely to be honoured and to be faced with the inevitable bickering which follows.

Should payment not be made within the prescribed time, the contractor's normal remedy should be the right to claim interest at, say, 2 per cent above bank rate (see condition 40.2 of MF/1). A prolonged failure by the employer to pay should entitle the contractor to stop work – see condition 40.3 of MF/1.

### Recovery of payments made (6)

Where payments are made in advance of delivery to site the two rights which an employer will usually seek to have included are:

1  A bond to be lodged for not less than the amount of the down payment. The making of the payment and the lodging of the bond should take place at the same time, and the contractor should check that the time limits for doing both are the same. Cases have been known in which the time for lodging the bond ran from acceptance of the contractor's tender, while the time for making the down payment ran from the signature of the formal contract.
2  That where progress payments are made during manufacture:

    • plant to the value of the payment made is identified, becomes the property of the purchaser and is marked as such
    • such plant remains, however, at the sole risk of the contractor and is insured by him accordingly.

See for example condition 40.1 and Special Condition 40.1 Progress Certificates of Payment in MF/1.

# Retention money

Reference has already been made to the principle that retention moneys should be considered by the employer as a contractual safeguard, not as a cheap form of finance. The fixing of the level of retention money should take this into account so that no higher amount is retained than is reasonably necessary. Where the works are completed and taken over in sections these retention moneys should be released on a sectional basis.

The higher cost to the contractor of retention moneys on many plant contracts lies in the 5 or 10 per cent retained during the defects liability period.

It is to the contractor's advantage, therefore, to press strongly for the release of the final retention after take over against, if necessary, a bank guarantee. Nor is it considered that the employer's contractual interests would be harmed by such action.

A particular problem has arisen with the provisions in the JCT form of contract relating to the setting up by the employer of a trust fund into which retention money is paid (clause 30.5.1). The objective is clear that the retention money is held by the employer as trustee for the contractor in a separate account and does not therefore on the liquidation of the employer belong to the liquidator. The problem in practice is that some employers delete the clause, others just do not set up the trust fund, while others seek to resist setting up the fund because they allege they have rights of set-off in excess of what would be the value of the fund.

While the contractor has the right to enforce the setting up of the fund, if necessary by mandatory injunction, this right is lost once the employer is actually in receivership or liquidation. As regards the employer's rights to refuse to set up the fund for, say, a right of set-off which he alleges to exist, the better opinion seems to be that the mere allegation of such a right would not be sufficient (*Concorde v Colgan* (1984) in the High Court in Hong Kong).

## Set-off

Unless the terms of a contract for the sale of goods or for work and materials expressly provide otherwise a debtor is always entitled to set-off against a claim by a creditor, a sum for damages which he claims are due to him from the creditor, for example for delay to the work. This right must be distinguished from the right of abatement, under which a purchaser may defend himself by showing that the work done is worth less than the amount claimed, because for example it is defective or work has been omitted. Legally the two are different, so that when the term 'set-off' is used in contract documentation it refers only to its legal meaning as defined by decisions of the courts, although it is doubted if the decision is well understood other than by lawyers – see *(Acsim (Southern) Ltd v Dancon Danish Contracting*, 19 CON LR 1.

The problem which arises in practice with set-off is that, say, a nominated sub-contractor seeks summary judgement against the main contractor for money certified by the architect as being due to him. The main contractor raises the defence of set-off, claiming there are defects and delays in its completion which have caused him loss and expense. If the defence is successful, summary judgement will be denied and the case must proceed to trial with all the consequent delays this will involve for the sub-contractor in receiving payment.

The courts in such a case do require something more than a mere allegation of delay and defects to support a claim for set-off, but the amount of detail and quantification needed is a matter for the court's discretion. This can reduce significantly the practical value to the unpaid contractor or sub-contractor of the right to apply for summary judgement. While it is only fair that a sub-contractor should not be able to obtain judgement for payment when his work is late or defective, it is equally unfair that the main contractor should be able to put forward spurious arguments which allow him to retain money in his own hands which properly belongs to the sub-contractor. Clauses 23 and 24 of the JCT nominated sub-contract form NSC/4 attempt to overcome this problem but inevitably the provisions are long and complicated.

Latham in his report (recommendation 25, para. 8.9) has recommended legislative action to outlaw certain practices in relation to set-off, which may be the only solution to this particular problem.

# 15 Time for completion

Completion on time is not something which just happens. It has to be planned and worked for, and this process starts from the initial definition of the employer's objective in relation to the contract. In the achievement of completion on time, the contract has three functions to perform:

- to act as a means of communications between employer and contractor
- to provide an incentive to the contractor to complete on time
- to give the employer an effective remedy against the contractor should delivery be delayed.

## Contract as means of communication

Proper communication between employer and contractor is one of the essential factors in successful contracting. But before the employer can communicate his requirements to the contractor he must have defined them for himself. Step number one, therefore, is for the employer to be quite clear in his own mind by what date he requires completion of the contract and what, for this purpose, completion means. According to the nature of the contract and the employer's purpose, completion may have one of a number of meanings, the most common of which are as follows:

- goods either ready for shipment or actually shipped on f.o.b., c.i.f. or other terms
- delivery of goods to the purchaser's store or construction site
- physical completion of the construction of the works on site
- plant and equipment commissioned and proved ready for commercial operation
- process plant passed its performance tests.

The choice of which definition to adopt will in part be determined by the type of contract and in part by the method of contracting which the employer has selected. Thus, if the employer decides to undertake for himself the actual importation of goods, he can hardly expect the contractor to be responsible for the date of their arrival in the UK. Equally if the contractor is only responsible for 'technical' supervision of erection he cannot be expected to guarantee the productivity of the employer's labour, or that the works are completed on time. It would be reasonable in such a case to require delivery of materials to be completed according to a defined programme with damages for delay attached to all key deliveries, and for the contractor to be responsible for any delays to the target date for completion of the plant as a whole which were caused by the negligence, incompetence or misconduct of his supervisors.

Where there is a contractual obligation in relation to delivery it is important to distinguish between actual delivery and readiness for shipment, particularly where the employer is arranging shipment, for example the delivery terms are f.o.b. From the contractor's viewpoint the control of the delivery operation up to the point at which the goods are ready for shipment lies within his own hands. After that, however, he is dependent upon shipping action being taken by the employer. Following the basic principles that one only accepts contractual responsibility for matters over which one has control, it is obviously preferable from the contractor's viewpoint, and indeed reasonable, for his obligation to relate to readiness for shipment.

One of the most common sources of misunderstanding regarding completion relates to mechanical/electrical/process plant, as between the physical completion of such plant, its readiness for commercial operation, including having passed its completion tests, and its having passed its performance tests. The time gap between these three can sometimes be many months, either because the plant, due to its nature, must be brought into operation slowly – for example, a brick kiln – or requires a running-up period, or because, due to the novelty of the plant or process being used, there are inevitable teething troubles. It is vital, therefore, that at the earliest moment possible in the 'thinking' of the project the term 'completion' should be defined, and that this definition be used in all subsequent planning.

The definition of 'completion' to be selected in this type of case will depend primarily on whether the employer or the contractor is responsible for the plant's performance or process design. If the process or functional design is that of the employer, then completion would normally be when the plant has been built and tested mechanically. If, however, the process or functional design was that of the contractor, then the plant would not normally be regarded as completed until it has been shown to be ready for commercial operation.

This is not of course the same as having passed its guarantee tests. Normally the plant will only have been run for a short period to demonstrate its ability to operate, usually referred to as a reliability run. The conditions for such a run must be defined within the contract and as a minimum should include:

- period of the run
- notice to be given of when the run is to commence
- list of permitted outages, i.e. those items which will not count against the contractor if they occur during the period of the run, such as malfunctioning of other related

equipment not under his contract, default of the purchaser etc.

- proportion of the period the equipment must function correctly and the maximum period of any single non-permitted outage
- who is responsible for operation and maintenance of the plant during the run
- records to be kept of the equipment's behaviour
- procedure for repeat of the run if this becomes necessary.

Although it is not provided for in the standard forms such as MF/1 the employer may of course require that the plant does pass the guarantee tests as part of the tests on completion. This is most likely to be the case when the plant is being financed on a project finance basis and the lenders insist that before the plant is taken over it has been proved to work to the guaranteed requirements.

It is essential, particularly where large sums of money may depend upon whether the plant was 'completed' on a certain day or not, for the criteria and mechanism for deciding this issue to be set out in the contract. The draft contract document at p. 162 does this by relating 'completion' to the contractor's right to apply for a Taking Over Certificate and there must be a test procedure for obtaining this certificate laid down elsewhere in the contract, probably in the specification. It must of course also be practical for the employer to have provided by that date facilities for the necessary testing to be carried out.

Another important factor to be considered is the relationship of the definition of completion to (1) the take over of the plant by the employer, that is in his assuming responsibility for accidents or damage to the plant and responsibility for its maintenance and security; and (2) payment by the employer of the whole or part of retention moneys.

With building and civil engineering contracts the problem is often that the purchaser is concerned with access to the whole or sections of the works before the final completion itself. Thus if final completion is made the only contractual obligation, the contractor could comply with this and yet, if late on the prior access dates, could cause the employer considerable financial loss. It is essential with this type of contract therefore to decide on the date or dates by which access is required, to make these firms contractual obligations, and to attach damages for delay to each.

The same principle applies to say a contract for a power station associated with a major process plant in which there are five sets of turbines to be brought into use sequentially over a period of time as the process plant becomes progressively operational. There needs to be a contractual completion date for each set.

Naturally the inclusion of such a provision for sectional completion is likely to lead both to an increase in the contractor's price and an extension of the contract period and also to a more rigid attitude by the contractor as regards his rights to extensions of time. However these disadvantages must be weighed by the purchaser against the fact that unless sectionalized completion dates are included he may find, as in the process plant example, that the works are 90 per cent completed and paid for (other than for retention moneys), yet are only 25 per cent operational and he has no contractual remedy.

Having settled on the definition of, and date or period for, completion the employer's next step must be to communicate that information to the contractor. This communication should be regarded as one of the essential items to be included in the inquiry or invitation to tender, or supplied as part of the data on which negotiations are to proceed in those cases where the contract is on a negotiated basis.

It is sometimes suggested that the delivery period should itself be made the subject of competition and the tenderers asked to quote their best offer. This can cause difficulties. Delivery is normally related closely to both specification and price. Decision on one will affect the other. Shorter delivery can be achieved in a variety of ways: by overtime and weekend working, by selecting those bought-in components which are themselves on shortest delivery, or by lowering standards in construction work on site. How is the tenderer to read the purchaser's mind? How is he to judge what price the purchaser is willing to pay for time?

If the employer really is interested in obtaining competitive offers on time, then it is suggested that he can do this in the following way. The basic inquiry against which all tenderers must quote includes a fixed date or period for completion. The tenderers, however, are also invited to quote as an alternative for an improved delivery and to give the following information regarding their offer:

- period by which they could shorten delivery
- additional cost for improved delivery per week or month as appropriate
- what methods they would use in order to obtain the improved delivery and any qualifications or understandings on which the improved delivery is based
- what guarantees they would be prepared to offer in respect of the improved delivery.

Adoption of this suggested approach would ensure, first, that all offers were obtained on the same basis and could therefore easily be compared and, second, that the employer has had all the information necessary to see whether it was practical to buy time and, if so, how much this would cost.

Having obtained a delivery promise in a tender which suits the programme, the purchaser is often inclined to think that such a promise holds good no matter how long the placing of the order/contract is delayed. This cannot normally be so. Any delivery promise is contingent upon the contractor's own work programme and the delivery periods being currently quoted for materials and bought-out components. These can all be subject to rapid change.

## Limit to value of undertaking

Ideally, it is suggested that the tender should be accepted or the contract should be placed within 30 days, or in the case of very major contracts up to three months. If this cannot be achieved, then it is no use just sending off the letter of acceptance quoting the original delivery promise and hoping for the best. It is often tempting for the contracts officer at this point, with the order in one hand, to press the contractor to undertake still to maintain his original promise to complete by a certain date despite the delays which have arisen since his tender was submitted. The contractor for his part, in his anxiety to secure the business, may easily be weak enough to give way to such pressures, only to regret it later when it becomes apparent that delays are inevitable. Any such temptation should be resisted. Not only is the practical value of an undertaking obtained in this way extremely limited but, worse, the contracts officer has allowed himself to be deluded into thinking that he has negotiated a favourable bargain. The planning of the contract and any associated work will proceed on the basis that the completion will be as promised when it almost certainly will not.

There is only one delivery promise worth having, and that is one which is as factual as it can be made and has reasonably taken into account the known sources of probable delay. For this reason, if the contract cannot be placed quickly, then the contractor should be given a reasonable opportunity to confirm the original promise. If it is a large contract, then it is often advisable at this point to discuss the programme with the contractor in order to make sure that nothing has been overlooked and that delivery terms from major sub-contractors or suppliers have been rechecked. It is helpful for any such discussions regarding confirmation of delivery promises to be held not merely with the contractor's sales staff but also with their planning, construction, or works people present. The normal pessimism of one is a good antidote to the over-optimism of the other.

In order to ensure that delivery promises included within a tender are realistic, the more information that can be given to the tenderers on the factors affecting delivery the better. Such information should include (depending on the nature of the contract):

- date or period after contract for access to site

- dates or periods for the supply by the employer of drawings or information
- dates or periods for completion by the employer or other contractors of work interrelated with the contract work
- restrictions on availability of site or working hours
- special inspection or approval procedures or quality standards demanded
- use of the site or common facilities by the employer or other contractors of the employer
- restrictions on spending of money within defined periods
- requirements as to completion of the work in a certain sequence and any dates for completion of sections of the contract
- dates or periods for the provision by the contractor of defined drawings or data.

Few contracts involving work on site can proceed independently of the employer or other contractors of the employer. Nor is it possible to make the best use of time and resources if the planning of the order and sequence of operations is left to one party. The employer may want certain sections completed before others; he may require from a plant contractor loading data and drawings for foundations design by a particular date. The civil contractor has to balance the most effective utilization of plant and labour and the relation of the workload of certain trades to anticipated programme and weather conditions. Plant contractors may require access to parts of a building in a certain sequence. Inevitably, all these interests will at some time conflict; also they may have a major effect on the contract price. This is why it is so important that the planning and coordination involved are to some extent worked out before, not after, the tendering stage and key points established and made clear in the tender documents.

Two objections may be raised to this suggestion. First, that it restricts both the contractor's initiative and that of the client's engineer. Second, that by making these times and periods part of the contract, if the employer should default on his obligations then he is laying himself open to a claim.

As to the first, the time has long since passed when either a single contractor or the employer himself can act independently. Projects are growing in complexity all the time, and this complexity in turn has led to the growth of the number of specialist suppliers and sub-contractors whose work is closely related one with another and with that of the main contractor.

Regarding the second, it must be accepted that once one introduces planning into a project the employer, no less than the contractor, becomes bound by the times and periods set out in the plan. If these have to be altered, then the person responsible for the alteration must bear some liability for the consequences.

## Critical path analysis

Any detailed discussion on the use of network analysis would be outside the scope and purpose of this book. The principles behind the technique are by now fairly well known, and those wishing to gain further knowledge of it are advised to consult one of the numerous specialist books on the subject. What is perhaps, however, not so well known or brought out by the books dealing with the technique are those factors which may limit its effective application and which need watching if it is to be of maximum benefit. These may be summarized briefly as follows:

1  The technique cannot of itself improve the nature of the data used. If this is inaccurate, then so will be the answer. The danger is that because the answer has been obtained from a network, perhaps with the aid of a computer, it will be assumed to possess a significance far greater than an answer obtained by simple, old-fashioned methods.

2  Because as a technique it is interesting and has attracted its own devotees, it is easy for it to be treated as something which has a justification to exist in its own right. It has not. It must prove itself to management by providing a quicker and more accurate answer to the problems affecting the control of a project than any other method, thereby enabling significant

economies to be made, if it is to survive. It remains at all times a tool of management and under management's control. Most definitely it must not be allowed to become the preserve of the analyst or programmer.

3   Following on from note 2 above, it is for management to lay down the manner in which it wants the output data presented and how this data is to be translated into effective instructions to the company's executives and site staff. It is very easy for this essential step to be overlooked. If it is, then it may be found that, instead of the network being used as a practical working aid, it is ignored by the very people, the project engineers and resident site staff, whom it was intended to benefit.

4   The other enemy of the network is detail. Because a network is easy to expand perhaps it is inevitable that it should be expanded. If, however, a project manager asks an engineer whether it will be safe to increase the load on a particular foundation he wants the answer, not a mass of calculations. The same thing is true of the network. He wants to know if the project is on time – if not, why not – and what can be done about it. If it is on time, then what is it essential to be doing next to ensure that it remains on time? These are simple questions which demand simple answers in plain English.

# Contract as means of providing an incentive

The contract may provide an incentive to a supplier/contractor to achieve completion either on or in advance of time in broadly one of three ways:

● by the method of payment of the contract price being such that any delay will cause the contractor additional expense, and correspondingly that earlier completion will save expense

● by offering a bonus for earlier completion with a corresponding 'penalty' for late completion

- by a profit-sharing arrangement under which the combined effect of savings in cost and time are shared between the purchaser and the contractor.

## Method or terms of payment

There are several ways in which the method or terms of payment can give the contractor an incentive to early completion:

1   Where the contract is on a lump sum basis for the carrying out of site work, the contractor's overheads will have been estimated on the assumption of the site work lasting so long. Any extension of that time will cost the contractor money.

2   Where payment of the contract price is at defined rates for units of completed work – for example, a yard of advance tunnelling – then unless progress is achieved to programme, the contractor will still have to pay out the costs for hire of plant, overheads, and wages of direct labour, but will not be recovering for these on the basis on which he prepared the estimate.

3   If the contract price or a proportion of it is withheld until completion is effected, then any delay will cost the contractor interest charges and lose him working capital.

## Bonus and penalty

The incentives to the contractor referred to above are in the negative form, in that failure will result in a loss. While this is of some effect, the carrot is often more effective than the stick. A positive inducement may therefore produce better results than the threat of being penalized. The difficulty is to make certain that the bonus really is effective in producing a greater than normal effort. Before offering a bonus, therefore, it is necessary to establish the norm both in time and price.

It follows that a bonus is something to negotiate after tenders have been received, not something to be mentioned when tenders are invited. There could be a difficulty under the Public

Procurement and Utilities Directives in doing this unless the employer was entitled to, and did, use the negotiated procedure. It is considered that if the restricted procedure has been used then any such negotiations would have to follow the contract award. Assuming that the procedure referred to earlier, of inviting tenderers to put forward alternative offers on time, has been followed, it may be found that the lowest tenderer has offered to complete a month earlier for a £50 000 increase in the contract price. If a month is worth more than £50 000 to the employer it might be reasonable to negotiate on the basis that for completion in a month earlier one would pay a bonus of £50 000, for completion on time no bonus or penalty and, for late completion, then damages at least equal to the amount of the bonus.

Particular care has to be taken when negotiating a bonus and penalty clause on a cost-reimbursement type of contract. The danger is that, to earn the bonus, the contractor will spend the employer's money to an unreasonable extent. It is necessary, therefore, to establish that the bonus and penalty are related not only to time but also to the excess of the actual costs over target. Thus a table (see Table 15.1) might be included in the contract on the following lines, with the bonus/penalty applied only to the contractor's fixed margin. The purchaser would continue to pay actual costs although, as stated earlier, depending on the degree of uncertainty, he might put a total limit on his liability.

It will be noted that Table 15.1 is worked out on the basis that:

- savings or increases of £100 000 on cost are worth 5 per cent of fixed margin
- one week of time is equal to 5 per cent of the fixed margin
- savings or losses in time or money are not expected to exceed £400 000 or 4 weeks.

While in terms of damages it is reasonable to grant extensions of time for delays outside the contractor's control, since to do otherwise would be both unfair and put up the price, the same considerations do not apply to the bonus. The employer is only interested in paying the bonus for results. It is suggested, therefore, that extensions of time in a bonus clause should only be allowed for delays due to acts or defaults of the employer. These must be allowed since, once having undertaken to pay the contractor a sum in a certain event, the employer must not act in such a manner as to deny the contractor the opportunity of so doing.

**Table 15.1 Cost table showing bonus/penalty**

| | | COMPLETION EARLY WEEKS | | | | COMPLETION ON TIME | COMPLETION LATE WEEKS | | | |
|---|---|---|---|---|---|---|---|---|---|---|
| | | 4 | 3 | 2 | 1 | | 1 | 2 | 3 | 4 |
| Costs | +400 000 | M | −5 | −10 | −15 | −20 | −25 | −30 | −35 | −40 |
| above | +300 000 | +5 | M | −5 | −10 | −15 | −20 | −25 | −30 | −35 |
| target | +200 000 | +10 | 5 | M | −5 | −10 | −15 | −20 | −25 | −30 |
| | +100 000 | +15 | 10 | 5 | M | −5 | −10 | −15 | −20 | −25 |
| Costs equal to target | | +20 | +15 | +10 | +5 | M | −5 | −10 | −15 | −20 |
| Costs | −100 000 | +25 | +20 | +15 | +10 | +5 | M | −5 | −10 | −15 |
| below | −200 000 | +30 | +25 | +20 | +15 | +10 | +5 | M | −5 | −10 |
| target | −300 000 | +35 | +30 | +25 | +20 | +15 | +10 | +5 | M | −5 |
| | −400 000 | +40 | +35 | +30 | +25 | +20 | +15 | +10 | +5 | M |

All figures percentages to be added to or subtracted from the fixed margin M as the signs indicate.

## Liquidated damages for delay

The principle behind liquidated damages for delay is that they should be an accurate pre-estimate of the losses which, at the time of entering into the contract, it is estimated the employer would be likely to suffer were completion to be delayed, and which would arise directly out of such delay. The amount of such loss, and therefore of the damages, may and very often does bear no relationship at all to the value of the contract. Yet in commercial practice it is almost universal for such damages to be expressed as a percentage of the contract price. The reason the damages are really there is not so much to provide the employer with an effective remedy, but to protect the contractor by establishing a limit to his liability.

Commercially, in fact, this must be so. No contractor can afford to be liable for a risk against which it is difficult to insure and which is out of all proportion to the value of the contract and to his anticipated profit. It is only, therefore, in a limited number of cases that there will be any direct relationship between damages for delay and potential loss. Provided that the damages are less than the estimated amount of the loss, this legally does not matter, but what must be understood is that, having taken his remedy by way of liquidated damages, the buyer cannot, because the actual losses exceed the estimate, seek to recover the difference.

This point is illustrated in an extreme way in a recent case under the JCT form of contract where the employer had included in the Appendix under the heading 'liquidated and ascertained damages nil'. When the contractor was late in completion the employer sought to argue that the intention had been to delete clause 24 (the liquidated damages clause) from the contract and he was therefore entitled to claim damages at large. The court ruled that the parties were free to include within their contract whatever figure they chose subject to it not being a penalty. Having agreed that the liquidated damages were nil that is exactly what they were to be; further having left the liquidated damages clause in the contract this excluded any right of the employer to claim damages at large.

Reference is made above to the term 'penalty'. The distinction between liquidated damages and a penalty is peculiar to English law. The difference was well stated by Lord Dunedin in the classic case of *Dunlop Pneumatic Tyre Co. Ltd v New Garage Motor Co. Ltd* in the House of Lords in 1915 AC 79.

The essence of a penalty is a payment of money stipulated as in terrorem of the offending party; the essence of liquidated damages is a genuine covenanted pre-estimate of damage . . . it will be held to be a penalty if the sum stipulated for is extravagant and unconscionable in amount in comparison with the greatest loss which could conceivably be proved to have followed from the breach.

It follows from this that if a single amount is payable under different circumstances in one of which only it might be reasonable pre-estimate of loss but in the other it could not possibly be that the payment will be held to be a penalty. In practice, this means that if the employer wishes to claim liquidated damages for say failure to complete the works on time and also for failure to provide handbooks and as-built drawings then he cannot seek to claim the same amount for both.

However, in other situations the courts have shown recently a welcome commercial approach towards the distinction between a penalty and liquidated damages. In *Phillips Hong Kong Ltd v Attorney General of Hong Kong* (9 February 1993) the Privy Council held the fact that a clause could in various hypothetical situations, none of which had actually happened, result in a larger sum being recovered than the actual loss suffered by the employer did not make the clause a penalty. The Privy Council also emphasized that when parties are of equal bargaining power then the court should be too ready to find the requisite degree of disproportion.

If the clause is held to be a penalty then the result is that it is unenforceable at law but of course the contractor still remains in breach of contract for being late. The employer's remedy is then to claim damages at large but these have, of course, to be proven, and what, from the employer's viewpoint is even worse, is that, as they are not liquidated, they cannot be deducted from the balance of the contract price.

Although the total value of the liquidated damages is unlikely, therefore, on any major contract to be equal to the employer's potential loss, the employer, by adjusting the rate at which damages are recovered, may be able to correct this under-recovery over a short period. Thus if 5 per cent of the contract price per week represents a fair pre-estimate of the loss, then instead of damages at the rate of 1 per cent to a maximum of 15 per cent as often applies in the case of electrical or mechanical plant contracts, the damages could be expressed as 5 per cent per week to a maximum of 10 per cent. The employer is here exchanging the high maximum for a recovery rate over a short period in line with his anticipated rate of loss. He may, however, find the contractor unwilling to accept such a rate unless he is allowed a 'grace' period before the damages start. Nevertheless the rate at which damages are to be recovered is something which wants to be kept completely flexible and tailored to suit each individual contract.

Other points which arise on the drafting of the delay in completion clause are as described below.

*Definition of the amount on which the damages are payable.*    This may be the contract price as a whole, the contract price of a section, if there are damages attached to the completion of sections of the work, or the contractor may suggest that damages be calculated only on that part of the plant which cannot in consequence of the delay be put to the use intended.

If in fact the employer can make use of a plant or building for the purpose for which it was intended even if a particular section is late, or there is late delivery of handbooks or spares, then it is considered that if the contract were to provide for damages to be payable on the whole of the works even although that section or item were late then this would be construed as a penalty and as such unenforceable at law. The same argument would apply if a plant were divided into, say, three sections which could be utilized independently and only one was late. This is because the employer would have taken the same remedy in damages for the happening of two different events – delay of the whole and delay of the section or item – which must have a different effect on the loss which he would suffer which is the basis of the liquidated damages assessment (at least in legal theory).

*The rate at which the damages accrue.* It makes a great deal of difference whether the damages are expressed to be payable '£. . . . . . . . per each full week of delay' or 'at the rate of . . . . . . . per week'. In the first case the contractor is granted six days' grace before any damages are payable at all; in the second case he must pay damages at one-seventh of the weekly rate from the first day.

The damages are often expressed to be in full satisfaction of the contractor's liability for delay. The first point to note on this provision is that in respect of contracts which are subject to the Unfair Contract Terms Act its enforceability is subject to the court being satisfied that it is 'reasonable' (see p. 321 et seq.).

Second, although the provision is included in most standard forms of contract and also in contracts which are individually drafted, the question arises as to just what it means and what is its legal effect. Assume that the clause states that the damages are to be at the rate of 1 per cent per week to a maximum of 10 per cent. Does this mean that the liability of the contractor for a ten-week delay is limited to 10 per cent, or that the contractor's liability for damages is limited to 10 per cent irrespective of the period of delay? The problem is discussed fully in *Electrical and Mechanical Engineering Contracts* by K F A Johnston (Gower, 1971).

The author's view is that subject to the Unfair Contract Terms Act there is no reason why the parties should not agree to limit the contractor's liability for delay to, say, 10 per cent regardless of the actual length of the delay involved, but that it would require very explicit wording to persuade an arbitrator or judge that such were in fact the parties' intentions. Further it would need to be specifically stated that such a limit was still to apply if the employer were to exercise his right of termination once the period related to the maximum of the liquidated damages had expired. It can certainly be argued that some meaning must be given to the phrase 'up to a maximum of $x$ per cent' since under the principles applicable to the concept of liquidated damages the employer cannot recover more for each week of delay than the percentage stated. In the absence of such explicit statement as that referred to above the opinion is preferred that the maximum limits the right of the employer to the recovery of liquidated damages but does not prevent him from exercising any other right which he has under the contract or otherwise. He could therefore, after the maximum period has expired, give notice to the contractor either terminating the contract or requiring it to be completed within a reasonable period of time. If the contractor were then to fail so to complete the employer could exercise both his right of termination and claim damages at common law for the period of the delay after the end of that to which the liquidated damages relate.

Since the last edition of this work the MF/1 Conditions have followed this principle, although they do require that the contractor be given notice to complete within a reasonable time once the maximum has been reached (clause 34.2). After the expiry of this notice the employer has the right either again to require the contractor to complete, or to terminate, and in either event to recover his losses up to the limit of liability expressed in the contract.

The following additional points relative to the subject of liquidated damages are worth noting. There is no truth in the old tale still sometimes told that a liquidated damages clause cannot be enforced unless a bonus is also agreed. The clause can also be enforced even if the actual loss suffered by the employer is less than the amount included in the contract; it is sufficient if the employer can show that it was reasonably foreseeable at the time of entering into the contract that he would suffer damages at least equal to those included in the contract.

The liquidated damages can even be recovered if, in the event, the employer has suffered no loss at all provided, as always, that it can be established that at the time of entering into the contract the level of damages agreed did not represent a penalty in the terms as described above.

*The contractor to be entitled in certain circumstances to an extension of time.* Sometimes an attempt is made to list the circumstances (see for example the condition in the JCT Form of Contract no. 25.4). More normally in UK contracts the phrase used is 'act or omission of the purchaser or the engineer or any industrial dispute or any cause beyond the contractor's reasonable control'. It is necessary to include industrial disputes separately, as otherwise it might be argued that an industrial dispute was within a contractor's reasonable control. A further point of significance in relation to the recovery of liquidated damages and clauses for extension of time is that where the employer is wholly or partially responsible for the contractor's failure to complete on time, the employer cannot recover liquidated damages unless the contract expressly provides otherwise. It is important to note that a general clause referring to 'force majeure or other unavoidable circumstances beyond the contractor's control' will not suffice to cover the employer's default. The result will then be that liquidated damages cannot be deducted and the contractor's obligations as regards completion will be to complete within a reasonable time (*Percy Bilton v GLC* [1982] 20 BLR 1). Any extension of time clause whether expressed in general or extensive terms should therefore always include specifically 'any act or omission of the employer or the engineer'.

One term which should not be used in an English contract unless its meaning is defined and it is only 'shorthand' for that meaning is 'force majeure'. The term is derived from French law and has no legally defined meaning in English law. Its use, unless its meaning is defined in the contract, can only lead to confusion. Contracts with overseas purchasers also often refer to the term 'force majeure' and careful check should be made as to the meaning which this has in the foreign legal system, since it can vary widely.

It is also advisable from the contractor's viewpoint that the word 'reasonable' should be included, as it is believed that this would enable the contractor to argue, for instance, that whereas it was within his control to overcome some difficulty if he spent a large and disproportionate sum of money, it was not within his reasonable control, as the use of the word 'reasonable' implies that financial considerations can be taken into account.

Some support is given to this contention by the case of *B&S Contracts & Design v Victor Green Publications* which was decided in the High Court in 1984. There it was held that an employer who acted 'unreasonably' in not paying money to prevent a strike of his workmen in circumstances in which he must have known that non-payment was likely to result in their going on strike and their demands were not exceptional was not protected by a force majeure clause in the contract. Presumably if the circumstances had been different and the payments required by the workmen had been out of all proportion to what was in the circumstances 'reasonable', then he could have claimed the benefit of the clause. It also implies that practical factors can be taken into account (see further on this point, extensions for time for delays due to sub-contractors, on p. 284).

It is also interesting to note that the High Court have recently decided (Times Law Report, 25 May 1994) that the expression 'reasonably practical' goes beyond what is physically feasible to include financial considerations. While the case was not concerned with the implementation of a contract but of a court order it again supports the contention given in the text above.

Legal systems other than those based on English law do not recognize the distinction between liquidated damages and penalties. In French law, for example, a penalty is enforceable and it is recognized that one of the purposes of the penalty clause is to encourage the contractor to complete on time. But the penalty represents the maximum of the amount for which the contractor is liable unless the court decides that it is derisory. However, in other systems such as German the contractor may also, if the purchaser can prove that he has suffered a greater loss, be responsible for the extra, i.e. the liquidated damages clause does not necessarily constitute a limit. If therefore the contractor is working under a foreign legal system the position needs to be carefully checked.

# 16 Sub-contracting

Modern industrial activity is based on specialization and the combining of specialist skills to form an integrated whole. Where this integration function is the responsibility of the contractor/manufacturer, it follows that a substantial proportion of the actual work will be sub-contracted or comprise bought-out items. On an industrial building contract, the actual work to be carried out by the builder's own labour may represent only a very small proportion of the total contract price. The remainder will be sub-contract trades – for example, plasterer, tiler, asphalter, and bought-in items of equipment and sub-contract work. It is clear, therefore, that in preparing the contract very careful consideration must be given both to the control which the employer has over sub-contractors and to the responsibility which the main contractor bears for them.

Normally the contract between the employer and the main contractor does not create any contractual rights or obligations as between the employer and the sub-contractor. If the goods which the sub-contractor supplies or the work he carries out prove defective, then the employer's remedy is against the main contractor.If on the other hand the main contractor fails to pay the sub-contractor for work done or goods supplied, then again, unless the contract specifically provides otherwise, the sub-contractor has no recourse against the employer.

If however a supplier of material guarantees to the employer that his product will be suitable for use on a particular contract, and the employer as a result specifies their use, then, as described on p. 53, the courts may decide that there is a collateral contract between the employer and the material supplier under which, in consideration of having his materials specified, the supplier guarantees their suitability.

## Employer's right to restrict sub-contracting

The employer has no authority to prevent the main contractor/supplier from sub-contracting parts of the contract work, unless the contractor has been selected and the contract placed on the basis, either express or implied, that the work is to be performed by the contractor himself. It is rare to find such a provision written into the contract expressly, and normally the only circumstances in which it will be implied is where the contract by its nature is one for the provision of personal services – for example employment of a particular consultant because of his special expertise.

For all practical purposes, therefore, a contractor/supplier is free to sub-contract any part of the contract work subject only to the express terms of the contract.

One question which may be asked at this stage is why the employer should wish to exercise control over the employment of sub-contractors. The reasons would seem to be as follows:

1   The employer knows and presumably approves of the standards of workmanship of the main contractor/supplier. He does not in all probability have the same knowledge of the sub-contractor, and although the contract would normally entitle him to have any defects remedied, this would inevitably involve the employer in trouble and expenses for which he would probably be unable to recover

in full. Prevention is better than cure.

2  Extensive employment of sub-contractors increases the difficulties of the main contractor in co-ordinating the work and may be evidence that he has over-reached his capacity in taking on the contract.

3  Where site work is involved, the employer may have reservations about the sub-contractor concerned entering on to his premises. Again a multiplicity of sub-contractors can cause labour difficulties.

These are all valid reasons, and no employer can afford to allow a main contractor unlimited freedom to sub-contract. At the same time, the exercise by the employer of this control does raise certain problems as follows:

1  Any control can be irksome and create delays unless exercised with flexibility and understanding.

2  While exercising his rights to object to a sub-contractor whom he considers to be unsatisfactory, the employer will normally wish to avoid getting into the position of accepting responsibility for the choice of sub-contractors.

3  The establishment of any direct relationship between the employer and a sub-contractor will lead to a weakening of the main contractor's own position as the 'employer' of the sub-contractor and could lead to the employer being regarded as having a responsibility towards the sub-contractor.

## Safeguards for employers

In order to provide the employer with reasonable safeguards and at the same time minimize these problems it is suggested that the following steps should be taken during the course of the contract negotiations:

1  The contract conditions should contain a prohibition against sub-contracting without the employer's or his engineer's

consent other than for the supply of materials or for minor items. According to the nature of the contract it may be worthwhile defining in more detail either any particular item about which the employer wants to be consulted – for example, the supplier of an unusual or difficult casting on which there have been previous problems – or those where the employer is prepared to leave it to the main contractor, perhaps all below a certain financial level.

2  The contract conditions should state expressly that the employer's consent to or approval of a sub-contractor does not relieve the main contractor of any of his obligation, and he remains fully responsible for the acts and defaults of the sub-contractors.

3  The invitation to tender should contain a schedule for completion by the tenderers of the work which they propose to sub-contract and the names of the sub-contractors whom they would intend to employ.

4  At the outset of the contract, the main contractor should be required to confirm his sub-contracting arrangements and to obtain the consent of the employer or his engineer to the employment of any sub-contractor not named in the tender.

The object behind requiring the main contractor to list the principal sub-contractors proposed in his tender and to put forward the names of the remainder at the outset of the contract is to remove any source of disagreement between the employer and the main contractor, if possible before the contract is awarded and at the least while there is still time for negotiation. If this is not done, and the employer does object to a particular sub-contractor, the main contractor may seek to argue that to alter the choice now will delay the contract and cause additional expense for which he has made no allowance in his price.

## Responsibilities of main contractor for sub-contractors

Ideally, from the buyer's point of view, the responsibility of the main contractor for all that his sub-contractors do or fail to do should be no different from that which the main contractor accepts in relation to work which he carries out directly with his own labour. The practice has, however, grown up in certain fields of contracting for the liability of the main contractor to be restricted and either:

- be no greater than the main contractor can himself impose on his sub-contractor or supplier, or
- be such that contractually the main contractor has no direct liability himself, but simply passes on to the employer the benefit of any warranties or guarantees offered by the sub-contractor.

In general this practice is inconsistent with the concept of a main contractor and denies to the employer one of the principle advantages of employing a main contractor, that of having only one firm responsible for the contract. As such it is clearly against the employer's interests. At the same time, it must be admitted that this practice has developed at least in part out of policies pursued by employers themselves. First, the more the employer seeks to control the selection of the sub-contractor, the less is a main contractor going to accept liability for the acts or defaults of the employer's nominee. Second, since acceptance of responsibility involves risks which must be allowed for by the main contractor when pricing the job, he is not likely to willingly accept such responsibilities unless the opportunity exists for him to include such 'cover' in his price. If, however, the buying policies of the employer are such that he insists that the main contractor only receives a small handling or procurement fee on sub-contracts, then the main contractor is denied that opportunity and accordingly is bound to seek to contract out of direct responsibility for such sub-contractors.

The employer may argue in reply that he is better off paying a low procurement fee and relying on the commercial pressure he can bring to bear on the sub-contractors, through the threat of their future business should they misbehave, than he is in having contractual rights against the main contractor and paying higher fees. Within a limited field where there are only a few companies placing business and these operate internationally – for example the oil and petrochemicals industry – there may be something in this argument, but it is clearly not of general validity.

There are certain occasions on which to seek to apply the principal of total liability of the main contractor would be unreasonable. Take the case where the contract includes a special item designed and manufactured only by one supplier. The main contractor cannot be expected to be an expert in the design of that item, and the risks involved may be out of all proportion to the sub-contract price. In this sort of example it would seem fair to make the main contractor's liabilities in respect of that item extend only to:

- the main contractor's own negligence or default – for example, supply of incorrect data or error in installation.
- the passing on to the employer of the best warranty terms which the main contractor can obtain from the supplier.

Much the same arguments apply to payment. The old saying 'he who pays the piper calls the tune' is as true as ever. The employer would be most unwise to pay the sub-contractors direct. The employer indeed has no authority to do so unless expressly authorized by the contract. Further as the law now stands if the main contractor were in liquidation the employer would run the risk of double payments, once to the sub-contractor and once to the liquidator.

With that background the contractual responsibilities of the main contractor for his sub-contractors may be considered under two broad headings: liability of sub-contractors for defects; and time.

## Liability of sub-contractors for defects

The main contractor should normally be fully responsible for defects caused by his sub-contractors and suppliers over the same guarantee or defects liability period as for his own work. The question which then arises is that of the liability of the sub-contractor to the main contractor. There are three issues here. First, is the sub-contractor to be liable for making good defects in the sub-contract works for the same period as that for which the main contractor is liable or, assuming the subcontract works are finished earlier, does his defects period run from the date of the completion of his own works?

Second, does the sub-contractor's statutory period of limitation run from the completion of the main contract or the completion of the sub-contract?

Third, does the sub-contractor have the obligation to indemnify the main contractor for loss or damage which the main contractor suffers under his contract with the employer, to the extent that this is due to the default of the sub-contractor?

The last issue can be dealt with easily. All the standard forms of sub-contract in the construction industry, including the form for use with MF/1, contain wide-ranging indemnities in favour of the main contractor. These can be extremely onerous on the sub-contractor especially since the period of limitation only begins to run from the time when the loss is established or incurred. It seems indeed doubtful if the full impact of these indemnity provisions is properly understood by many sub-contractors.

Practice as regards the first issue varies. The MF/1 sub-contract form clearly specifies that the sub-contractor's defects liability period is the same as that of the main contractor. Other forms, such as the nominated form NSC/4 for use with JCT80, provide that the sub-contractor's period of defects liability runs from the practical completion of his own work.

As regards the second issue, at the time of writing the position is that generally the period runs from the completion of the sub-contractor's work. But this solution is controversial and there are proposals supported broadly by clients, designers and main contractors that there should be a single period of limitation of liability running from the completion of the main contract – that is, the limitation period should be project-based. This is objected to by sub-contractors, who consider that the present position that there are separate limitation periods for each sub-contract should be maintained. In practice, having regard to the sub-contract indemnity clauses, it is difficult to see what real advantage the sub-contractors gain from their opposition.

One problem which sub-contractors genuinely have, especially those who are specialist firms, is that as a purchaser of materials for incorporation in their work they could well find their suppliers declining to accept any liability at all beyond a six-month period for proven defects. Assuming the parties to be of equivalent bargaining strength such clauses might well survive an attack under the Unfair Contract Terms Act 1977.

These issues are widely discussed in the Latham Report and in *Product Liability in the Construction Industry* by Palmer and McKendrick, Lloyds of London Press, published in association with the Joint Contracts Tribunal. It seems however that they are unlikely to be resolved in the near future.

In the meantime, to judge from the evidence collected in the Latham Enquiry, there is clearly a wide level of dissatisfaction felt by sub-contractors in their relationships with main contractors especially in the areas of contract conditions and finance.

That this is the case is not altogether surprising. While it may be prudent from the main contractor's viewpoint to seek extensive indemnities, and to operate on what is largely a 'back-to-back' basis with their sub-contractors, this ignores the commercial reality that the main contractor is paid to take the overall risks of the project and to supervise the work of his sub-contractors. One can draw the conclusion that if main contractors paid more attention to these aspects of their work, and less to trying to protect themselves against their sub-contractor's default, then perhaps there would be less dissatisfaction in the industry.

It is also important from the employer's viewpoint that he does nothing which would undermine the main contractor's position in his relations with the sub-contractor.

In negotiations where the employer is claiming against the main contractor, due to a defect in a specialist sub-contract item, it may often appear that the main contractor is acting as no more than a post office and the employer may be tempted to take matters into his own hands and deal with the sub-contractor or supplier of the specialist item direct. This is a temptation which the employer in his own interests should resist. Once an employer has direct contact with the sub-contractor he not only makes the main contractor's position impossible, but may easily prejudice any contractual rights which he has against the main contractor.

## Time for completion

In the negotiations of fixed completion periods and so-called 'penalty' clauses, two of the principal objections put forward by contractors against the acceptance of such contractual obligations are, first, that they cannot impose like terms on their suppliers and sub-contractors and, second, that they may be delayed in the completion of the contract by the failure of the sub-contractor in circumstances in which it was impossible for them to compel the sub-contractor to complete on time.

As to the first objection, it is again part of the main contractor's job to organize his sub-contracting in the most effective manner possible and to place his sub-contracts on the most favourable terms he can. Even, however, if the sub-contractor does accept a fixed completion period and damages for delay, the level of those damages will almost always be related to the sub-contract price, thus leaving a gap between the main contractor's liability to the employer and what he can recover from the sub-contractor (see, for example, clause 7.1 of the MF/1 form of sub-contract). There is really no wholly satisfactory answer to this problem; it is part of the main contractor's risk for which he earns his margin on the sub-contractor's price.

One partial solution which has been tried is to insist that the sub-contractor in default bears the whole damages payable up to the limit contained in his sub-contract. An example may make this clear.

- value of the main contract: £1 000 000
- value of the sub-contract: £100 000
- damages under the main contract: 1/2 per cent per week to a limit of 5 per cent

- damages under the sub-contract: 5 per cent per week to a maximum of 10 per cent
- the contract was four weeks late due to the sub-contractor's default.

The main contractor would be liable therefore to a total of damages of £5000 × 4 = £20 000. Of this the sub-contractor would indemnify him to a total of 10 per cent of £100 000 = £10 000, leaving the main contractor to find the other £10 000. If, however, the delay was only two weeks, then the sub-contractor would be liable for the whole of the damages.

The second objection has more validity, and it is suggested that the commercial basis for the recovery of damages should be not only that the main contractor is late, but also that he has in some way defaulted in his own obligations. These may be expressed in relation to sub-contracting as follows:

- selection of reliable sub-contractor and obtaining the employer's consent to their employment as required by the contract
- placing of the sub-contract at the appropriate time to fit the overall project programme, having obtained a realistic completion promise from the sub-contractor which fits the programme
- actively progressing the sub-contract from its commencement
- advising the employer at the time it occurs of any delay which is likely to affect the overall programme and taking all reasonable steps to overcome that delay.

### When extension of time is allowed

If notwithstanding the above the job is still late, due solely to the sub-contractor's default, then provided the contract contemplated that work being sub-contracted, that is, that it was work included in the contract but of a type not normally carried out by the main contractor himself, it is thought that the main contractor ought to be entitled to an extension of time. Support for this proposition is to be found in the House of Lords' decision in *Scott Lithgow v Secretary of Defence* 1989. There the contract for two submarines contained the words 'In the event of exceptional dislocation and delay arising from . . . *any other cause beyond the contractor's control*' and went on to provide for the effect being assessed by the parties or for the Ministry to pay for the vessel on an 'actual cost basis'.

Delays were caused because of manufacturing defects in the special cables supplied by BICC. In holding that the contractors were entitled to the benefit of the clause Lord Keith stated

Prima facie it is not within the power of a contractor to prevent quality breaches of contract on the part of a supplier or sub-contractor such as lead to delay. The contractor has no means in the ordinary case of supervising the manufacturing procedures of his supplier. He specifies his requirements but has no means of ensuring that they are met. . .

However, MF/1 clause 33.2 provides that a delay by a sub-contractor which prevents the contractor from completing on time will entitle the contractor to an extension of time, provided that the delay is due to a cause for which the contractor himself would have been entitled to an extension. This appears to imply that in any other circumstance, i.e. if the sub-contractor is simply in default despite the main contractor's best efforts, then the main contractor is liable to the employer for the resultant delay and is left to whatever remedy he has against the sub-contractor. Support for this proposition is provided by the case of *Fairclough Building Ltd v Rhuddland Borough Council*, 5 October 1983 where the standard JCT contract had been amended to state that Fairclough were entitled to an extension of time for delay by a nominated sub-contractor 'which they had taken all avoidable steps to delay or reduce but such delay will be only considered for these reasons for which the main contractor (Fairclough) could obtain an extension of time under this contract'. Since on the facts the sub-contractor's delay was due simply to their own default Fairclough were not entitled to any extension of time under their contract although equally there was no evidence of any default by Fairclough themselves.

If there are no specific provisions in the contract conditions then it is considered on the strength of the *Scott Lithgow* decision that a main contractor could claim an extension of time if he could establish that the default of the sub-contractor was in practical terms beyond his control. This would obviously not apply if it was work which the main contractor could reasonably have been expected directly to supervise, but perhaps only to specialist manufacturing work in the sub-contractor's works.

# Nominated sub-contractors

Stress has so far been laid on the desirability of the main contractor having the widest choice of sub-contractor possible, consistent with the employer retaining technical and commercial control of the contract. There is, however, a practice which is particularly prevalent in building and civil engineering contracts, under which sub-contractors and suppliers are nominated by the architect/engineer. Very briefly, the system is that at the tender stage a prime cost or provisional sum, representing the estimated value of the work, is inserted by the architect/engineer in the bill of quantities, and when a sub-contractor has been selected by the engineer/architect the main contractor is told to place his sub-contract with that firm. In practice, certain nominated sub-contractors may be selected before a decision is taken on the main contract. The estimated amount in the bill is then replaced by the actual sub-contract price. The main contractor, for his services, is paid a fee on the sub-contract price for profit and attendance.

The system is convenient as regards fittings, for example locks, doors, sanitary ware, etc., in that it saves the architect having to specify these in detail at the tender stage and allows him time to choose those he considers most appropriate. Applied, however, to large sub-contracts for building work such as structural steel, heating and ventilating or electrical work, it has many disadvantages as follows:

1   It removes these sections of work from the competition for the main contract.

2   No matter what the contract says, the main contractor never feels the same degree of responsibility for a nominated sub-contractor as for one of his own choice.

3   The employer has to assume responsibility for the integration of the programmes of the nominated sub-contractors with the main building programme. It follows that very often no really firm programme can be established when the main building contract is placed.

4   The system has worked against the growth of medium-sized construction firms who are capable of tendering for and handling integrated contracts for a complete project inclusive of steelwork, mechanical and electrical services.

5   Under the procedures for nomination in both the JCT and ICE contract forms the courts have effectively placed the risk on the employer of repudiation by a nominated sub-contractor or justified forfeiture of the sub-contract by the main contractor for the sub-contractor's default.

6   Since in general the employer's rights in respect of defective work by the nominated sub-contractor can be exercised only through the main contractor, who will normally have played no part in the selection of the nominated sub-contractor or in the writing of the terms upon which the nominated sub-contractor tendered, it is necessary in the main contract conditions to cover in some detail both how the rights of the employer are to be preserved and how the main contractor is himself to be protected. Suppose for example that the nominated sub-contractor insists on contracting only on the MF/1 conditions which as is noted in many instances – e.g. limitation on liquidated damages, responsibility for making good defects, liability for accidents and damage – are substantially more favourable to the contractor than the ICE conditions. Is the main contractor to be obliged to accept the additional liabilities which he cannot pass on, or are the employer's rights as regards the work covered by the nominated sub-contract to be limited to those which the nominated sub-contractor is willing to accept? This problem and others are dealt with in some detail in clause 59 of the ICE conditions and generally in a manner which seems fair

to both parties, but necessarily the provisions are complex and much care is needed in carrying out the procedures involved if the intent of the various sub-sections of the clause is to be realized. So much is this so that one wonders if the alleged benefits of nomination are worth the effort involved and the risk, if any of the procedures are not properly followed, of a break in the contractual chain which leaves the employer without adequate remedy or of the contractor being saddled with responsibilities without the possibility of enforcing them.

One issue which has been tackled in the latest edition of the ICE conditions is the problem of where the works to be performed by the nominated sub-contractor include design whilst design is not under the ICE form part of the main contractor's responsibility. It is now provided that if the design requirement is specifically included both in the main and nominated sub-contract then the contractor is liable to the employer for such requirement (clause 58 (3)). However, under clause 59(1) the main contractor can object to the employment of a nominated sub-contractor who declines to enter into a sub-contract under which he accepts towards the main contractor like obligations and liabilities to those which the main contractor accepts towards the employer. Since it may fairly be said that the ICE conditions are not really designed for contracts under which the contractor assumes a design liability, it is likely that any commercially prudent nominated sub-contractor for M&E work would so decline and insist on the use of MF/1 or a like set of conditions. The engineer will then be obliged to proceed under clause 59(2) and either nominate another sub-contractor or, which is more likely, omit the works from the contract. There is no longer any provision allowing the engineer to nominate on terms not complying with clause 59(1).

An important change was introduced in the 6th edition of the ICE conditions to the effect that the main contractor is as fully liable for a nominated sub-contractor as for a domestic one, unless the default of the sub-contractor gives the main contractor the right to terminate the sub-contract. In those circumstances the main contractor is indemnified by the employer for his losses and expenses which he cannot recover from the defaulting sub-contractor.

There is also a new provision 59(1)(d) that a ground of objection to a nominated sub-contractor is that he will not provide the main contractor with security for the proper performance of his contract. It is not clear whether this refers to an 'on-demand' bond or not, or what level of security the main contractor can require, but presumably he could insist at least on the same type and level of bond as he has been required to provide himself.

The position is even more complex under the JCT conditions of which a significant part is solely concerned with issues relating to nominated sub-contractors. Within the scope of this work only three will be considered: those relating to delay in completion, defects in the work and that of re-nomination. For a more detailed commentary on the forms the reader is referred to *Keating on Building Contracts*, Sweet & Maxwell, 1991.

## Delay in completion

Under the main contract JCT 80 clause 25.4, the main contractor is entitled to an extension of time if he is delayed in completion of the works by reason of 'delay on the part of nominated sub-contractors or nominated suppliers which the contractor has taken all practicable steps to avoid or reduce'. This extension of time is not dependent upon the cause for which the nominated sub-contractor/supplier is delayed and extends even to default by him in the carrying out of his work.

The employer's remedy is through the direct warranty which he should obtain by using form TNS/1 for nominated suppliers and NSC/2 for nominated sub-contractors.

## Defects in the work

While a nominated sub-contractor or supplier is still a sub-contractor or supplier of the main contractor the latter's responsibilities which would otherwise exist for the sub/contract or supplier's work are substantially reduced by the express terms of the JCT80 contract. In summary the position is:

1   The main contractor is under no liability for the sub-contractor's or supplier's design, any performance specification forming part of the sub-contract or the suitability for purpose of materials which the sub-contractor or supplier supplies (clause 35.21 and clause 36). See also *Young & Marten Ltd v McManus Childs Ltd* [1969] 9 BLR 77, which had already established the general position that a contractor is not responsible for the fitness for purpose of materials which are specified by the employee or his architect.

2   The main contractor is responsible to the employer for the nominated sub-contractor's and supplier's workmanship and quality of the materials which are supplied. This was again established in the *Young & Marten* case. Accordingly he has the normal obligations of inspection and a liability for defects which such examination should have revealed. This obligation, subject to what is said below regarding restrictions in the sub-contract terms, means that the main contractor is also liable for latent defects in the materials or workmanship in the same way as he is for the remainder of the works.

3   If the sub-contract contains provisions limiting the liability of any sub-sub-contractor or supplier with whom the sub-contractor is required to contract which are approved in writing by the main contractor and the architect, then the liability of the sub-contractor to the main contractor and of the main contractor to the employer is similarly limited in respect of the sub-contract works (clause 35.22).

4   If the terms of contract determined by the architect with the supplier do not contain any provisions which limit the supplier's liability, the main contractor will be liable for latent defects in the materials supplied. If such terms do contain limitations on the supplier's liability, the main contractor's liability to the employer will be similarly limited *provided that the main contractor has obtained the approval of the architect/contract administrator in writing to those restrictions*

(clause 36.5).

From the contractor's viewpoint therefore it is essential that he notifies the architect of any term excluding or limiting the supplier's liability and obtains his approval before placing his order. Failure by the architect to give approval would entitle the contractor to reject the nomination.

What appears not yet to have become before the courts is what the position would be if the restrictions on liability insisted upon by the nominated supplier were held not to satisfy the test of reasonableness under the Unfair Contract Terms Act (see p. 321). Could the question be raised in an action by the employer against the main contractor? In principle there seems to be no reason why it should not be since the effect of clause 36.5.1 is to incorporate the supplier's terms into the main contract.

## Collateral warranties

Reference was made earlier (see p. 58) to the need for the employer to obtain collateral warranties in order to give him a direct right in contract against a defaulting sub-contractor. A collateral warranty is simply an agreement made between the sub-contractor and the employer in which the sub-contractor undertakes to the employer that he will perform all obligations contained in his sub-contract with the contractor. Further, to the extent that the sub-contractor is responsible for design, that his design will be fit for the purposes required by the employer. It is recommended that the warranty should be phrased in this way so that the sub-contractor's liability for his design is strict and not limited to the exercise by the sub-contractor of reasonable skill and care.

The consideration for the collateral warranty is usually expressed as a nominal amount of money, although if the collateral warranty is executed as a deed then strictly consideration is not required.

The circumstances in which a collateral warranty should be obtained can be summarized as:

- on any contract where the sub-contractor is undertaking

specialist work which is critical to the functioning of the works

- where a nominated sub-contractor is responsible for the design of any part of the works
- on any management contract where the design liability of the contractor is limited to the damages which he can recover from the defaulting sub-contractor. In this instance it may be necessary to bond the sub-contractor's liability under the collateral warranty, since the most likely reason for the inability of the management contractor to recover damages is that the sub-contractor is in receivership.

There are further circumstances where other parties such as financiers and future purchasers of a development may require collateral warranties but these are outside the scope of this book. For details of these reference may usefully be made to *Collateral Warranties*, Frances A. Patterson, RIBA Publications Ltd 1991, with January 1993 Supplement.

## Re-nomination

In the leading case of *North West Metropolitan Hospital Board v T.A. Bickerton & Son Ltd* [1970] 1 WLR 607 it was held that if a nominated sub-contractor fails to complete his work then the employer is under a duty to re-nominate and it is the employer who must bear the increased costs of completion by the new sub-contractor and some part of the main contractor's losses caused by the delay. The primary reason behind the decision was that the main contractor was barred under the terms of the contract from carrying out the work himself and therefore it must be implied that there had to be a re-nomination. This position has more recently been confirmed by the Court of Appeal in *Fairclough Ltd v Rhuddlan Borough Council* where it was additionally made clear that the re-nomination had to cover not only uncompleted work but also work which the original nominated sub-contractor had done imperfectly from which it followed also that the employer was responsible for the costs of remedying the work done improperly. Further, in order to be valid the nomination had to provide for the work to be done and to be performed within the original overall contract completion period, or the main contractor given an appropriate extension of time, objection having been made by the main contractor to the proposed re-nomination on the grounds that the sub-contractor's completion date was beyond that of the main contract (see *Building Law Monthly*, October 1985).

# Review

It is evident now that the way in which the nominated sub-contractor system has developed, under the standard building and civil engineering forms in particular, is that the main contractor has become substantially a co-ordinator and that the employer as regards his rights for defective design and lack of suitability for purpose of materials is largely looking to the separate agreements and warranties concluded between himself and the sub-contractor/supplier. Especially on building work under the JCT forms there is not one main contract but a complex series of inter-locking agreements between the employer and the several nominated sub-contractors and suppliers. The system, given its obvious difficulties and the substantial burden of administration work which it imposes, is now increasingly lacking in support. The CIPS in their submission to Latham described it as 'a contradiction in terms' and recommended its abolition. It is reported that only 11 per cent of specialist engineering contractors are nominated under JCT80. Latham himself did not recommend that it should be followed as a normal procedure. It is hoped that the days of the system are numbered and it has been abandoned in the NEC. But there is also evidence to suggest that in protest against this complexity the use of the somewhat simpler JCT Intermediate and Minor Building Work contract forms is becoming more widespread and being applied to a higher value of work than was ever originally intended.

There is an argument that the system results in lower prices than would be the case if the main contractor had to accept the entirety of the risks involved. This is probably true to the extent that through the nominated system the employer has taken away from the main contractor his power of choice, but it is also likely that the out-turn costs will be higher because of problems of co-ordination especially in the engineering design.

There are a number of alternative routes that can be taken in lieu of the present nominated system:

1   The employer could seek to utilize the rather simplistic mechanism of the government conditions of contract, GC/Works 1, 1991 edition (see clauses 61 and 62). Effectively these make the main contractor liable for the nominated sub-contractor as if he were a domestic sub-contractor, except for the risk of determination of the sub-contract through liquidation. In respect of design carried out by the nominated sub-contractor, the main contractor is responsible in the same manner as a professional architect/engineer would have been had he carried out the design under an independent contract with the employer (see clause 9).

2   The architect, in conjunction with other specialist designers, could provide performance specifications for work such as mechanical and electrical services against which the main contractor would bid as part of his tender, selecting his own domestic sub-contractor from a short list given in the enquiry documents. The sub-contractor would undertake the necessary detailed design work for tendering purposes and the main contractor would be required to state in his tender the sub-contractor he had chosen and with whom he would be required to sub-contract the design and execution of the electrical and mechanical works. In this way continuity of design and installation would be maintained. Any time lost in the tendering process as compared to the nomination system would be recovered post-contract.

3   The employer could place separate contracts for the main elements of work which would normally be undertaken by nominated sub-contractors and through the construction management system would co-ordinate these. Each contractor would be directly responsible to the employer.

4   The employer could appoint a specialist firm as main

contractor and allow him to select his own civil or building contractor as a sub-contractor. This is only feasible if the specialist firm has the capability of acting as main contractor and exercising effective control over the civil or building firm.

Whichever method is adopted two objectives need to be achieved. There must be a clear responsibility for design and for the integration of the design of the various sub-systems and the building work. The employer must have clear and simply expressed contractual rights in respect of any default by a specialist firm either through the main contractor or directly against the specialist firm itself without the need for a proliferation of collateral warranties.

Conc?

# 17 Delivery

Under a simple order for the supply of goods, the supplier's total liability for the goods will normally terminate when they leave his factory. After that the extent of his contractual responsibility will vary according to the specific terms of the contract. If he is responsible for making delivery to the buyer's store, then he must arrange carriage of the goods either in his own transport or under a proper contract of carriage with a third party; and, unless he obtains the buyer's specific consent, such contract should not be at owner's risk. In so far, however, as it is the purchaser who has the duty of taking delivery of goods and inspecting them on arrival, the supplier will insist that, if he is to be liable for any loss or damage during transit, that notice is given by the purchaser in time for the supplier to comply with the carrier's terms of contract. Whether or not he is to be so liable will depend on the terms of the individual order, but, from the buyer's point of view, the only safe course to adopt is to assume that if it is desired to make the supplier take the risk of the goods in transit, then this should be expressly written into the contract. To rely in this instance on implied terms is to tread on dangerous ground.

## Responsibility during installation

These arrangements are reasonable enough where the contract is completed (other than for the provisions of the defects liability clause) when goods complying with terms of the order and specification have passed into the buyer's physical possession. The position is rather different, however, when one is concerned with plant which has to be assembled or installed on the buyer's premises or construction site, and then commissioned by the supplier before he can be said to have fulfilled his contractual obligations. In this case the purchaser is not really concerned with the individual units making up the plant but with the whole, assembled, tested and in proper working order. It might be suggested therefore that the contractor in such event should retain the ownership of the goods and the absolute liability for them, until the point has been reached when the plant is taken over by the purchaser, when the property and risk should pass. This would indeed be a simple solution, but it is not in many instances a practical one for the following reasons:

1   The contractor will usually want to be paid a substantial percentage – say 90 per cent – of the value of the goods when they are delivered. Having paid all but the retention money for the goods, the purchaser will naturally want them then to become his property, so that he has security for the money paid.

2   In very many cases it would be uneconomic to require the contractor physically to take delivery of, and arrange storage for, the various units of the plant as they are delivered to the construction site or the buyer's premises. It would mean the establishment by the contractor of a site organization which at least in the early stages of the contract would only be employed part-time. It is commonly arranged, therefore, as part of the services which the purchaser is to provide under the contract, that the purchaser will be responsible for taking delivery and storing the parts of the plant until they are needed. In so far, however, as the purchaser performs these tasks he cannot at the time expect the contractor to take the

legal responsibility if anything should happen to the goods whilst under the purchaser's physical control. If it is intended that the purchaser should be responsible for reception and unloading of the plant and its storage, condition 24.1 of MF/1 would need to be amended accordingly.

It is normal, therefore, for conditions of contract governing the supply and installation of plant to provide that the items of plant making up the works become the property of the purchaser when either they are delivered to site, or the contractor becomes entitled to require their value to be included within an application for a certificate for payment. For a typical example of such a clause see condition 37 of MF/1.

While, however, the purchaser is happy to become the legal owner of the plant, he is not so anxious to assume complete responsibility for any loss or damage to it which may occur at any time up to the plant being taken over. It is usual again, therefore, to provide that, except for any period when the purchaser is actually handling or storing the plant, it remains at the risk of the contractor. This must, however, be said expressly, since otherwise the purchaser, as the legal owner of the plant, may well find himself treated as the person upon whom the liability for any loss or damage may fall under the old common law concept that risk and property in the goods go together.

It is particularly important that if the purchaser is providing any services – for example the use of a crane for unloading, storage accommodation and the like – the respective responsibilities of the parties in this regard are made absolutely clear at the tender stage, and in this connection the following checklist may be found useful:

1  Is the purchaser or supplier to take delivery?
2  Who is providing labour and tackle for unloading?
3  If the purchaser is providing tackle, who is in charge of the operation and who accepts the risk if any accident occurs?
4  Is the purchaser providing storage accommodation or merely storage space for the contractor to put up his own store?

5   If the purchaser is providing storage accommodation, does he accept responsibility for superficial examination of packages and the like on delivery and for giving notification of any apparent damage or shortfall?

6   If the purchaser is providing storage accommodation, does he also accept responsibility for safe custody of the goods and for the suitability of the storage accommodation provided and methods of storage? Is it necessary for the supplier to advise on any special requirements – for example, for electronic equipment?

## Access to site

A further point to be considered is that of access to the site. Unless the contract states otherwise, it is the responsibility of the purchaser to provide access to the site of the nature which will permit the proper execution of the contract in the manner contemplated. MF/1 accordingly provides that:

The Purchaser shall provide such roads and other means of access to the Site as may be stated in the Specification subject to such limitations as to use as may be imposed (clause 11.1)

... approaches ... to be provided by the Purchaser shall be provided within the time specified in the Contract or in the Programme, shall be of the quality specified and in a condition suitable for the efficient transport, reception ... of the Works (clause 11.4).

If, therefore, there are any peculiarities regarding the site or restrictions on access which would interfere with normal delivery or make it more difficult – for example a bridge capable of only carrying a limited load – the purchaser, to protect himself against misunderstandings and ultimately a claim for extra payment, must set out the position expressly in the specification accompanying the invitation to tender.

The delivery of materials to site in order to ensure that they are not there prematurely also requires attention. Many sites are congested; storage space is limited, and there are often a number of contractors each wanting their allocation of the room available. Moreover, the risk of loss or damage, particularly to expensive or delicate items, is obviously far greater on a construction site than in the manufacturer's works, and while contractually the supplier may be liable to replace or repair the damaged items, the time taken to do so may have a serious effect on the programme for the project as a whole. Unfortunately, two factors combine to provide the contractor with a substantial incentive to make, or press his vendors to make, early delivery of materials to site.

The first is the fear of erection/installation work being held up by material shortages. The combination of repeated late deliveries from vendors and escalating costs of site labour has made this into a very real fear. Second, under the system of payment included within most standard forms of contract, under which the contractor is expected to finance the job in the early stages and is only paid for materials delivered to or work done on site, the contract itself provides the contractor with a built-in incentive to deliver and ensure that his suppliers deliver early, so as to get paid early. The incentive is all the greater in a time of high interest charges and restrictions on the availability of capital.

Two suggestions are made which it is thought might help to alleviate these problems. First, the main contractor should carry out more intensive programming of deliveries and expediting of vendors, including the expediting of the major supplier's sub-vendors. This latter point is known to be controversial, but some large contractors do it because they find that they cannot rely on their suppliers to carry out expediting of their own sub-vendors effectively. It is the old problem of the one specialist item holding up work to the value of many thousands of pounds, and only by the most intensive and integrated action can this be prevented or the effect of it minimized. In terms of pure contractual procedure, such action by the main contractor may be wrong, and it is agreed that it could lead to a blurring of responsibilities as between the main contractor and his suppliers. But the fact remains that suppliers will not accept, and indeed cannot reasonably be expected to accept, responsibility for the effect which their delay has on the whole contract. So the main contractor must look after his own interest, and in any event prevention is better than cure.

Second, payment on plant contract should not be tied wholly to deliveries made to or work done on site, but should be related to progress made against the contract programme. This point has already been referred to earlier under 'terms of payment' (see Chapter 14).

## Responsibility during storage and defects liability period

Despite all efforts to the contrary it not infrequently happens that the purchaser is not ready either for delivery of plant to be made or for its erection or installation to proceed. His own programme may be behind; related building or civil engineering work may be late. Arrangements must be made therefore for the items of plant to be stored and for the contract to be adjusted in such a way that, while the plant contractor is not penalized for something which is not his fault, the employer's interests also are safeguarded. The following points accordingly arise:

- responsibility and payment for storage
- liability for the plant while it is in store
- payment for the plant while it is in store
- effect of delay on the price for erection/installation
- carrying out of delayed acceptance tests
- adjustment of the defects liability period.

The responsibility for either storing the plant or arranging its storage should be placed firmly with the contractor, unless it has already been delivered to site, when this may no longer be practical. In that event the employer will have to accept the storage responsibility, for it is suggested that he would be wise to insist on the contractor preparing the plant for storage, inspecting it periodically during storage and advising on any special method of storage which may be needed. For any of these services the contractor will of course be entitled to additional payment. The contractor himself cannot, however, reasonably be expected to accept the obligation to store indefinitely. MF/1 now provides a more elaborate procedure than the old Model form A for dealing with delayed delivery which is the purchaser's fault, in that the effect of such delay is to suspend the progress of the works to the extent that progress is dependent upon the delivery of the plant delayed. However, the principles remain that it is for the contractor to store the delayed plant and that after a time period, now reduced to 90 days, the contractor has the right to require an instruction to proceed. If this is not received, he is entitled either to require a variation order to omit the work in question, to terminate (if the suspension affects the whole of the works) or to be paid the contract value of the plant affected by the suspension.

The effect of these provisions is to put considerable pressure on the purchaser to ensure that the other work which is necessary to allow either for delivery to be made or erection to proceed is completed on time. Also these provisions must be borne in mind by the purchaser when tendering and negotiating for contracts for the other works on which progress is dependent to ensure that the programmes are properly co-ordinated and that he is protected if the delay is due to the default of the other contractor(s).

If the plant is to be stored personally by the contractor, then he should be prepared to accept complete liability for any loss or damage which may occur in storage (other than any caused by an uninsurable risk), and he should be required to insure accordingly. The costs of such insurance would be payable by the employer. If on the other hand the contractor has no facilities to store the plant and must arrange storage with a third party, he is unlikely to be able to do so on terms under which the third party accepts such full liability. In that case it would seem reasonable for the contractor's own liability to the employer to be limited to whatever terms the contractor can obtain from the third party. There remains only the question of natural deterioration of goods during storage, which applies particularly to such items as certain electronic equipment. Obviously unavoidable natural deterioration is a risk which the employer must accept; other deterioration may be avoidable if expensive precautions are taken. Here it is for the employer to decide how much he is prepared to pay for, and for risks to be shared between contractor and employer accordingly.

Assuming that the terms of contract are such that the contractor only becomes entitled to payment for plant as it is delivered on site, it would clearly be most unfair if payment were to be withheld until the employer was ready for actual delivery to be made. The normal arrangement, therefore, is that on the plant going into store the contractor is entitled to be paid the same percentage of the contract price as he would have been entitled to receive on delivery being made to site. If the plant continues in store for a substantial period (under MF/1 conditions 120 days), the contractor is entitled to be paid such further proportion of the contract price as he would have been entitled to be paid on the issue of the taking over certificate. But, and this is most important, the taking over certificate itself is not issued.

It is always preferable under a contract for the supply and installation of plant for the erection or installation price to be shown separately in the contract. It is particularly useful when delivery has to be delayed, since clearly (*a*) this part of the price does not become payable until the work is actually carried out, and (*b*) if the delay is of any significance the contractor is bound to require such part of the price to be adjusted to take account of increases in wages or other costs which have occurred. This will be so even though the contract was originally on a fixed price basis, since such fixed price can only relate to the period of the contract as originally envisaged, and owing to the delay this may well mean that the impact of wage awards or government regulations on the erection price will be quite different from that estimated at the tender stage. There is the further point that, as a result of the plant having gone into store, some additional work may be necessary to put it into a condition to be installed. If so, then provided that this was not due to the contractor's default in any way, the additional costs should be added to the contract price.

It is most important to the purchaser to try to ensure that, despite the delay, his rights and remedies against the contractor in the event of the plant not being satisfactory are not unduly prejudiced. For this purpose two points must be covered: first, that the defects liability period does not start to run until the installation has been completed and the plant is actually taken over, and, second, that the contractor is still obliged to carry out the acceptance tests.

The contractor on the other hand cannot be expected to continue his obligations under the contract indefinitely. The solution contained in MF/1 would seem to be fair. This is as follows:

1 The defects liability period does not start to run until the take over certificate is issued.
2 The take over certificate is not in fact issued until the works have actually been completed.
3 The contractor is obliged to carry out the acceptance tests at any time during the defects liability period.
4 If delivery or installation of plant becomes delayed due to the actions of the employer or a person for whom the employer is responsible so that clause 25.6 applies, and the contractor is obliged to carry out his obligations under the defects liability clause more than three years after the normal delivery date for such plant, any additional costs incurred by the contractor shall be added to the contract price.

The issues discussed in this chapter apply with even greater force when the contract is being performed overseas. Replacement of goods lost or damaged takes longer and is more costly and the overall effect on programme and project costs is therefore that much worse. Thus it is even more important that the contract is clear as to where the responsibility lies for the performance of delivery in all its aspects. Additionally there are the following items peculiar to export contracts which need covering expressly within the contract:

1   Definition of the terms used such as f.o.b. or c.i.f. It is suggested that this is done by reference to the current edition of *Incoterms*, published by the International Chamber of Commerce.

2   Issue of the export licence if one is required – normally the responsibility of the contractor.

3   Issue of an import licence. The obligation should be that of the purchaser but he will require data normally in the form of pro-forma invoices from the contractor before he can act. A timetable for these events should be set out in the contract and preferably the contract should not come into force until the import licence has been issued.

4   Customs clearance. If the purchaser is a foreign government or quasi-government body then preferably this should be made his responsibility and he should be given a specific time within which to achieve it. Again however the contractor will be responsible for supplying the correct documentation in the required language and the requirements in this respect should be stated within the contract documentation.

5   Port delays. If port delays are anticipated the contract should provide that the completion date is based on a period of so many days between notification of a vessel's arrival and its ability to discharge cargo and any delays beyond this entitles the contractor to claim an extension of time.

6   Payment of duty. Government or quasi-government contracts abroad are often duty-free but only if the goods are correctly consigned to the purchaser and the cases carry the appropriate markings. This again needs to be specified in detail within the contract.

7   Method of transport. This may be dictated by the purchaser who requires the use of his own or a specified shipping/air line and often the use of particular agents. In this event the contract must provide a shipping period which if exceeded will allow the contractor to claim for delays. The procedure of any purchaser-appointed agents should be checked to ensure that they can be complied with within the proposed contract programme.

If the contractor is allowed the choice, then, assuming all three methods are available; land, sea or air, the primary factors to be taken into account contractually are:

- Safety and security of the goods. Air or a containerized load by truck or sea have a definite advantage on this account if circumstances allow.
- Availability of import control and customs clearance facilities. Many countries operate on the basis that the goods can only be cleared through the place where the import licence is physically held. It will be no use therefore deciding to transport a particular consignment by air in order to save time unless parallel arrangements are made to have the licence at the airport.

  Clearance as duty-free because it is a government contract may only be effected at certain entry points and these need to be identified.
- Restrictions on internal transport. Checks should be made on the size and weight of proposed loads against local roads, bridges and tunnels. Also, if internal air transport is restricted to the local air line, the capacity of its transport planes should be checked.

# 18 Defects: guarantees and remedies

Every purchaser would like the goods which he purchases whether commercially or privately to be perfect. But perfection is not something which just happens; it has to be worked for and paid for, often in terms of both cash and time. The higher the quality which is required, in general the greater will be the initial cost and the longer the delivery period. All this may seem axiomatic, but it is highly relevant to the methods of purchasing to be employed and the remedies which it is reasonable for purchasers to seek, against contractors who have apparently defaulted in their contractual obligations.

With every additional complex part which is added to any item and each extra processing operation which is included within the process of manufacture or plant operation, the probability of error arising is multiplied. It may be desirable, in the interests of advancing technical knowledge generally, to keep on with experimentation and to push even further forward with the development of new ideas. But there is a very distinct danger that the 'best can become the enemy of the good'. To set a time-scale on development is never easy; to utilize what is existing and available may seem dull compared with the excitement of further potential developments. But the balance must be kept between, on the one hand, falling behind technically and failing to take advantage of what can be achieved by bold experimentation and applying modern technology; and on the other hand never quite completing any development and achieving commercial success with it, before that development itself becomes outdated.

The buyer in his approach, in the specification he establishes and the guarantees he demands, sets the stage on which the contractor must perform. It is the buyer who fixes the priorities. Is it time that is vital, so that existing ideas and methods only can be incorporated? Is it a high degree of reliability, thus limiting again both design and production methods?

This is the buyer's decision. He will often want advice from the contractor on the time-scales and costs involved and the results which the contractor is prepared to guarantee as compared with those for which he will accept no contractual liability. What is vital is that the buyer recognizes the need for him to take this decision, and that he should frame his contract in accordance with the decision reached.

In addition to considering the above, the purchaser must also ensure that the contract correctly reflects the precise nature and quality of what he really needs. Over- or under-design can be equally expensive. There is no point in purchasing a high-quality article if, for the usage to which it will be subject, that quality is unnecessarily high. The same holds good the other way round. But the purchaser cannot have it both ways. Having accepted that the lower quality or lower performance, and therefore lower-priced, article or plant is suitable to his needs, he cannot then expect the same guarantees as if he had purchased the more expensive. If a processing plant has been designed to handle 100 tons of material an hour and is guaranteed at that figure, it is no use the purchaser complaining, after he has overloaded the plant by 25 per cent, that it has been inadequately designed. If he wanted a 25 per cent overload factor to be incorporated in the design, he should have said so expressly.

# Guarantees for materials, workmanship and design

## Guarantee period

The contractor usually wants to know that his contractual liabilities are clearly limited in terms of time and that this time is relatively short. In so far as materials and workmanship are concerned this is perfectly reasonable. With proper inspection, and after the plant has been in use for even a limited time, any defects due to defective materials or workmanship should have been revealed. Also, to the extent that the contractor has no control over the plant or the manner of its use once it has passed into the purchaser's possession, if the defects liability period were prolonged innumerable disputes could arise as to whether the failure was due to a defect in the goods when they were purchased, or whether it was due to subsequent mal-use or mis-operation.

The period which has been commonly accepted within the construction industry has been 12 months, but it is doubtful whether this period is any longer appropriate. It has been estimated that between 75 and 80 per cent of defects become apparent in construction work generally within the first five years from completion. It can reasonably be assumed therefore that a substantial proportion of defects only become apparent after the first year. Reasons why defects in the contractor's work can arise after the expiry of 12 months from completion include:

1  The technical complexity and novelty of the processes and materials now being used in construction and of the installed equipment.
2  The increasing extent to which the contractor who, although under a civil or building contract may not be responsible as such for design, is involved in practice in the 'design' of the works through the choice of materials and methods of construction. The dividing line between design proper and workmanship is often not easy to draw.

A significantly longer period than 12 months may now be

required for many contracts and this is a point which should always be considered by the purchaser when inviting tenders. Three further problems arise. First, from when does the period start? With plant purchased and taken into the purchaser's store, normally from the date of delivery; with plant purchased and installed, from the date when it is installed and tested ready for commercial operation (see p. 257). In either case, if for any reason there is a long delay in putting the plant into use, this can mean that the major portion of the defects liability period will have expired before there has been an opportunity of putting it to the test. In the same way as was suggested therefore in the previous chapter when dealing with delays in installation, the defects liability clause should cover this by giving the purchaser the right to advise the contractor that the goods will be going into store or not being used, and that in such event the period does not run until the goods are in fact put into use, depending on the nature of the goods. It should also be provided that the supplier should advise on any special methods of storage or protection required, and have the opportunity of inspecting the goods both during storage and when they are finally taken out; the contractor to make good the effects of any deterioration in the goods due to long storage or non-use, but at the purchaser's expense.

The position is more difficult when the plant concerned is not being stored but, after being installed, cannot be put into use for some considerable time due to other equipment not being ready, so that in its installed position the plant may be exposed to damage or contamination by dirt. An example would be plant installed in a ship's engine-room, which cannot be operationally run until the ship as a whole is ready for commissioning trials. In this instance, despite all precautions which may be taken on the contractor's advice and his inspection prior to the trials being held, if the contractor agrees to an extended guarantee period he is increasing his risk. He must be expected, therefore, to want a provision in his contract price additional to his normal allowance for defects liability.

With building or civil engineering work the period starts from the date of practical completion or substantial completion.

Difficulties have arisen with the term 'practical completion' in building contracts. As the JCT80 contract is written it would appear that the architect should not issue the certificate if there are any patent defects in the works unless these are very minor. In practice when employers are anxious to take possession this strict rule is not followed and the certificate is issued with a long 'snagging' list. However, when for financial reasons a commercial developer does not wish to take the building over the rule is strictly applied.

This lack of clear definition and variations in practice according to the employer's circumstances is clearly unsatisfactory, given the importance which attaches to the issue of the certificate of practical completion.

The second point is where a defect has been remedied within the defects liability period by the replacement of some part, as to what the liability period should be in respect of the part so replaced. Is it a further twelve months, or merely the balance then unexpired of the original twelve months? Many standard conditions of contract do provide for the former, and this is obviously to be preferred from the buyer's point of view, but equally, in fairness to the contractor, there must be a long-stop, say perhaps twice the original defects period. Additionally in some conditions of contract the defects period for the section of plant affected is extended by the time during which that section has been out of operation due to the defect. If the whole plant is put out of operation the period for the whole plant is extended.

Neither of these provisions is unreasonable, provided that there is a final limit to the defects liability period. No contractor should be expected to continue under a contractual responsibility indefinitely.

While, however, a fixed period of, say, 12 or more probably 24 months may be reasonable in relation to defects in workmanship and materials, it is doubtful if in certain circumstances this is satisfactory to the buyer in terms of design or specification where these are the responsibility of the contractor. The difficulty arises in this way. It frequently happens that a plant or item of equipment with a specific designed performance is not in fact operated continuously to that level of performance for some substantial time after it has been purchased or first put into operation. It may be run intermittently or with a much lighter load. The parts are not subject therefore to continuous running at the specified duty, and so defects in design which might otherwise have manifested themselves will remain hidden. Sometimes this problem can be overcome by making the guarantee in terms of design related to a specified number of hours' full load running. Alternatively the guarantee may be framed as $y$ months from the time when the plant starts continuous commercial operation at not less than $x$ per cent of its designed capacity, with an extension for any period during which it is out of operation due to a defect for which the supplier/contractor is responsible. In both cases the contractor would probably insist on a final maximum time limit from when the plant was put into operation so as not to leave his liability completely open-ended; and this would seem fair.

The other problem relating to design and the period for defects liability is that during only twelve months' operation even with normal usage a defect may well remain undetected, only to become noticeable some time later. This can happen also with civil engineering work such as foundations or dams; adverse conditions against which the design was supposed to have provided may not arise until after the twelve months' period has expired. Is the buyer in this sort of case to be left without any contractual remedy?

The answer should be 'no', provided that it can be established that the loss or damage concerned is due to a breach by the contractor of his warranty that the works as designed by him would be fit for the purpose intended. But the longer the time gap, the more difficult this is going to be; the use may have changed, unforeseeable circumstances may have arisen, and it must be remembered that the technical standards against which the design is to be judged are those which were prevailing at the time when the design was made.

Nevertheless, as was stated recently in the House of Lords by Lord Edmund Davies in a case involving a contractor's liability for a design failure: 'justice requires that we put ourselves in the position of [the contractor] when first confronted by their daunting task, lacking all empirical knowledge and adequate expert advice in dealing with the many problems awaiting solution. But those very handicaps created a clear duty to think through such problems so that the dimensions of venturing into the unknown could be adequately assessed'. In other words, the nearer the design is to the then 'state of the art' the greater the responsibility of the designer, more particularly if any failure would result in the likelihood of personal injury. As his Lordship further stated in the same case 'the law requires even pioneers to be prudent'.

As to what is a reasonable period after takeover to bring the contractor's contractual liability to make good defects due to a design fault to an end it is suggested that between three and five years could be appropriate depending on the nature of the works concerned. The position as regards the continuing liability of the contractor to pay damages as opposed to remedying the defect is considered later (see p. 325).

The third point which arises is whether or not any certificate issued by the architect/engineer at the end of the defects liability period operates as conclusive evidence that the works have been carried out in accordance with the contract and operates as a bar to any future legal proceedings in respect of latent defects.

Here practice differs as between the plant industry, building and civil engineering. Broadly with mechanical, electrical and process plant the final certificate does operate as a bar except in the cases of fraud and now under MF/1 36.10, except for latent defects due to the contractor's gross misconduct appearing within three years after take over.

In building contracts the latest Court of Appeal decision is that, under the JCT80 form of contract and also the IFC84 form, the architect's final certificate is conclusive evidence in any matter where the architect must be satisfied as to the quality of the materials and the standards of workmanship and that these conform in his opinion to the criteria required by the contract. The conclusiveness of the certificate is not limited therefore to those matters which the contract expressly provides must be to the architect's satisfaction.

Under the ICE conditions however the situation is the reverse.

Clause 61(2) provides expressly that the issue of the defects correction certificate at the end of the defects liability period has no effect on the rights of the parties in relation to the performance of the contract.

A distinction between plant and building/civil contracts is perhaps understandable because of the extent of liability to which the plant contractor could otherwise be exposed. The limitations in this clause and elsewhere on the plant contractor's liabilities have been vigorously defended – see for example the comments in the *Guide to the Use of the FIDIC Conditions for E & M Works*, 3rd edition, FIDIC 1988. What is strange and difficult to understand is the distinction in English law between building and civil contracts.

## Remedies available

When a plant or unit of plant breaks down the purchaser inevitably suffers losses. These may be broadly listed as follows:

- cost of replacement parts
- cost of stripping down and reassembly of replacement parts
- cost of repairing damage to other parts or units of the equipment or other property of the purchaser which may have been damaged
- damages payable to persons injured or owners of other property damaged as a direct result of the breakdown
- costs incurred in making temporary arrangements to overcome the effects of the breakdown
- loss of profits or increased overhead costs which are due directly to the breakdown
- damages which may be payable to a third party for breach of contract arising out of the breakdown.

Provided that the costs involved arise directly out of the defect and were reasonably foreseeable by the contractor at the time the contract was made, then in the absence of any express provision in the contract to the contrary all these items could form the subject of a claim by the buyer for breach of contract. But how far in commercial practice can the buyer reasonably expect to recover all or any of these costs from the supplier? Before this question can be answered the factors which may affect the contractor's attitude need to be stated. First, the contractor will broadly only accept a liability which bears some reasonable relationship to the degree of profit which he can expect to make out of the transaction. Second, the contractor has to take into account not just the risks on any one contract, but the sum of the risks on all the contracts of a similar nature into which he has entered and under which he has at any one time a potential liability. This is particularly relevant in the case of mass-produced articles, where the losses in which the contractor could be involved due to the failure of a single component could be astronomical. Third, the contractor is bound to consider on a swings-and-roundabouts basis the general level of his business with the particular customer. If he can assume £100 000 worth of business in any one year, then he may on a particular contract for £10 000 be prepared to take risks which are out of proportion to the value of that one contract.

Returning to the list of costs and expenses in which the purchaser may be involved, these may be divided into three categories (see Table 18.1).

Some conditions of contract prepared by suppliers or their trade associations exclude the supplier from liability for item 2, the labour charges involved in stripping down and reassembly. This would seem unreasonable. The same is true of carriage charges for the return of the defective part. If the machine is defective it should be the duty of the contractor to put that defect right, and he should be responsible for meeting all labour, material and carriage charges involved.

What is open to argument is how far the contractor's liability should extend over and above putting the defect right. Many firms take the view that anything beyond that is what they term 'consequential liability' and as such unacceptable. Apart from the two points already mentioned, of the relationship of risk to profit and the extent of total risk on annual turnover, contractors have other fears. First, they are afraid of claims being made

**Table 18.1   Costs and expenses in which purchaser may be involved**

| | | |
|---|---|---|
| 1 Cost of providing replacement parts. | Costs of repairing damage to other property belonging to the buyer caused by the defect. | Loss of profits or contribution to overheads arising out of the defect. |
| 2 Cost of stripping down and reassembly. | Damages payable to persons injured as a result of a defect.<br><br>Damage to property belonging to a person other than the purchaser. | Damages which may be payable by the purchaser to third parties for breach of contract as a result of the defect. |
| 3 ———————— | Cost of repairing damage caused to other parts of the machine or plant which the supplier has supplied or installed as part of his contract.<br>Costs incurred by the purchaser in making temporary arrangements to continue operations in order to overcome the effect of the defect. | |

which will occupy a disproportionate amount of their executives' time, and which it may be difficult to resist in the end due to commercial pressures. Perhaps even more important, they fear that they would have to dispute liability in many cases where, if it were simply a question of 'putting something right', they would concede and get on with the job, and that this could operate to the prejudice, therefore, of normal buyer/seller relationships.

For these reasons the problem of consequential liabilities needs to be broken down so as to arrive at a sensible sharing of risks between contractor and purchaser under arrangements which will:

● provide the purchaser with reasonable protection
● avoid the contractor inflating his price to cover against the risks or buying expensive insurance at the purchaser's expense
● minimize the chance of protracted disputes on liability

which are only likely to profit both companies' professional advisers.

It is suggested, therefore, that in the first instance a distinction should be drawn between before and after take over. Up to take over the contractor can reasonably assume the risks in the second column. After take over, when the plant will normally be insured by the employer and under his operation and control, it is preferable that they should be borne by the employer with the exception of the liability for death or injury to persons due to the contractor's negligence where the contract is one to which the Unfair Contract Terms Act applies.

If in a specific case the employer feels he must insist on these risks being taken by the contractor, perhaps due to political or trade union pressures, then the contract should:

● expressly define the liabilities to be covered
● include a clearly stated financial limit of liability, with a cross-indemnity by the employer for amounts in excess of that liability
● require the contractor to insure the risks up to the limits of liability and permit him to charge the premium in his price.

This, of course, still does not cover the items listed in the third column which from a commercial or indeed insurance viewpoint may be regarded as truly 'consequential' losses. No contractor is normally willing to accept these risks at all. They are too indefinite in amount and could be financially crippling. What is sometimes attempted is to include in the contract what is referred to as an 'outage' guarantee. This can take one of two forms:

1 If during the defects liability period the plant is out of operation for more than $x$ days continuously due to a defect, the contractor pays a fixed sum per day of plant outage.
2 If during the defects liability period the plant is out of operation for more than $x$ hours of operation, the contractor pays a fixed sum an hour of outage.

The employer can make out quite a reasonable case for this type of guarantee. He has spent a lot of money in the expectation of achieving a certain output and therefore a certain level of profitability. Repeated or extended shut-downs will seriously affect this. But there are practical difficulties involved.

First, there is seldom a single simple cause why a plant is out of operation. More often it is a combination of causes, some due to defects, some due to mal-operation or lack of maintenance. Pressure to keep a plant running, for instance, may lead to minor troubles being made substantially worse before corrective action is taken. While it is easy to write in to the contract that the contractor is not liable if the outage is due to the employer's fault, it is much more difficult to apply this in practice.

If the employer is going to insist on this type of provision, then the only real answer seems to be to let the contractor operate the plant. If this is impracticable, then at least the contractor should be made part of the operating team, say by being allowed to have an operating engineer on each shift, paid for by the employer, whose advice the employer is required to seek if difficulties arise.

Another problem is from what moment the period for outage liability starts. Statistics for many types of plant show a high incidence of minor troubles in the first few weeks, or hundreds of hours of operation. There must, therefore, be a sensible running-in period before the guarantee applies.

Finally, the extent of development included within the plant must be taken into account. An outage guarantee can only feasibly be given when the plant or design has already been substantially proven in commercial operation.

## Civil and building works

Attention has been focused so far on mechanical, electrical and process plant contracting where the issues involved are generally more complex. However especially with commercial buildings the employer will be likely to suffer so-called consequential damages if the defects in the building are such that it cannot for a period be put to the use to which it was intended. In the same way serious defects in civil works may mean a loss of revenue where these are related to a profit-earning project, e.g. a toll bridge.

It has never been the practice in civil or building contracts for the standard conditions to exclude the contractor from liabilities of this nature, so that in principle he could be liable to the employer under the normal rules relating to the recovery of damages for breach of contract (see earlier p. 60). It is also clear under these conditions that the contractor's liability to make good defects does not replace his common-law liability to pay so-called consequential damages subject to the tests referred to earlier of foreseeability and remoteness.

### Limitation and exclusion of liability

Although the express terms of a contract may seek to limit or exclude the liability of a contractor in respect of defects in the works, these terms may be subject to the provisions of the Unfair Contract Terms Act 1977. At the time when the last edition of this work was published it was also necessary to consider the extent to which such a clause could protect a sub-contractor or supplier if the employer were to bring an action against him in tort. It now appears that such an action would be most unlikely to succeed, first because it would be for economic loss (see p. 54) and second because 'there is generally no assumption of responsibility by the sub-contractor or supplier direct to the building owner, the parties having so structured their relationship that is inconsistent with any such assumption of responsibility' (per Lord Goff in *Henderson v Myrett Syndicate Ltd* [1994] 3 WLR 761).

# Unfair Contract Terms Act 1977

This Act made three important changes in the law so far as engineering-type contracts are concerned. First, it provides that a person cannot by reference to any contract terms exclude his liability for death or personal injury by negligence. Second, it provides that in respect of any other loss or damage a person cannot exclude his liability for loss or damage due to negligence except in so far as he can show that the contract term satisfies the test of reasonableness. Third, it provides that where a person is dealing 'on his own written standard terms of business' he cannot claim to render a contractual performance substantially different from that which was expected of him.

By seeking to restrict the contractor's liability to the express provisions of the defects liability clause the contractor is seeking to limit the purchaser's right to claim against him in two respects:

- the type of claim which can be made and
- the period during which a claim can be made.

Although there have now been a number of cases under the Act before the courts they are all in reality exercises in judicial discretion based on the particular facts relating to the case in question. For that reason it is difficult to be precise about what type of provision is likely to pass the test and what is not. All that can be done is to indicate some broad general guidelines as follows:

1   A clause which limits liability is more likely to be held reasonable than one under which liability is excluded.
2   If claims have been regularly settled by the contractor on a basis more favourable than that provided by the clause then this will be evidence that the clause is unreasonable.
3   The relative bargaining position of the parties. The stronger the bargaining position of the party seeking to invoke the clause the more likely it is to be held unreasonable.
4   The language in which the clause is framed and the size of the print used! The more obscure the clause the less likely it is to be upheld.
5   Which of the two parties it was more reasonable to expect to insure against the risk.
6   Whether there were any other options open to the purchaser, e.g. to contract on, to him, a more favourable basis at a higher price, to go elsewhere for an alternative source of supply on different terms.

7  As between commercial entities of equal bargaining power, that the clause was intended to be an agreed division of risk the significance of which was well understood by the parties who could be expected to be properly advised.

8  A clause purporting to exclude the contractor's liability for failing to comply with the express provisions of the contract relating to the performance of the works is unlikely to be regarded as reasonable, but it may be reasonable for the contractor to exclude consequential damages.

## Type of claim

Applying the above guidelines then in general it is suggested that it would be considered unreasonable to attempt to exclude any items listed in either the first or second columns in Table 18.1. It is more difficult to predict the court's attitude towards the items in the third column but it is thought that they would be less likely to expect the contractor to accept these risks.

There is again a difference between civil engineering and electrical/mechanical engineering conditions of contract in respect of the purchaser's right to the recovery of damages for defective work where the purchaser suffers loss or damage additional to the costs of remedying the defect. It would appear to be strongly arguable that the purchaser's right to recover such damages, which would include the loss suffered by the purchaser of not being able to make use of the works during the time taken to remedy the defect, is not removed because of the inclusion in the contract of the maintenance clause (clause 49 of the ICE conditions). In MF/1 however, clause 36.9, it is clear that all liability for any damage or loss attributable to the defect is excluded, although it seems that if the defect when it occurs causes damage to other parts of the works, the making good of that damage is the liability of the contractor. The question arises then as to whether such exclusion under the MF/1 conditions would satisfy the requirements of 'reasonableness'.

The MF/1 conditions, together with their predecessor Model form A conditions, unlike those of a trade association, are drafted under the aegis of professional bodies in an attempt to strike a fair balance between the interests of contractor and purchaser. Although in a particular instance the exclusion of the purchaser's right to the recovery of damages might appear to be harsh as it did to His Honour Judge David Smout QC in the case of *Southern Water Authority v Lewis & Duvivier and Others* 1 CON LR 40 – this must be balanced against the benefits which he otherwise derived from the conditions as a whole. The conditions in approximately their present form have been in use since 1948 and in respect of these exclusion clauses have never to the author's knowledge been subject to judicial criticism except as mentioned above in the Southern Water case where it was stated by the learned judge that if literally interpreted clause 30 (vii) of Model form A 'exceeded the bounds of commonsense'. The FIDIC conditions which are largely the same as MF/1 have been the subject of academic criticism – see the article by Andrew Pike in the October 1991 issue of the *International Construction Law Review*. He was strongly supported in his criticisms by Duncan Wallace QC in a letter to the editor of that Review in the July 1993 issue. In the course of that criticism Mr Duncan Wallace did raise the question as to whether or not in a domestic contract the provisions of clause 36 and others relating to limitation or exclusion of liability after take over would survive an attack under the Unfair Contract Terms Act.

The principle behind MF/1, FIDIC and the standard conditions of contract for process plants is the same. After take over the contractor is responsible for remedying defects during the defects liability period to the exclusion of any other liability for defects and once he has done this and the defects liability period has ended, in the absence of fraud, the contractor is free of liability. That principle has been widely accepted on both sides of industry for some forty years as creating a sensible balance between the interests of the contractor and of the purchaser, taking into account that it is the purchaser who is better placed to insure the risks arising after take over than the contractor.

Despite, therefore, the views of Mr Duncan Wallace, it is the author's contention that the MF/1 provisions would be considered as fair and reasonable were they to be attacked under the Unfair Contract Terms Act. However that criticism apart the fact that both sides of industry including on the purchasing side many public authorities with substantial bargaining power, have willingly been prepared over a large number of years to contract on these terms knowing their effect would appear to be a strong argument for saying that they are fair and reasonable.

## Period in which claim can be made

It has been noted already that there is a marked difference in contracting practice between mechanical and electrical engineering contracts as to the contractor's position at the end of the defects liability period. MF/1 clauses 36.9 and 39.12 make it clear that except in the case of fraud the final certificate issued at the end of the defects liability period is conclusive whereas the reverse is the case under the ICE conditions which provide in clause 61 (2) that the maintenance certificate is not to be taken as relieving either party from any liability towards the other arising out of the performance of their respective obligations under the contract.

So far therefore as civil engineering contracts are concerned, since the time limits within which actions for damages can be brought are those established by law, there would appear to be no room for the application of the Act. However with electrical and mechanical engineering contracts if the attempt is made to limit the contractor's liability to 12 months it would appear by no means certain that the courts would accept that such a provision was reasonable, at least in relation to defects of the type which are unlikely to manifest themselves during this period, e.g. design defects or defects related to a number of hours of plant operation under full load conditions which is not always feasible to achieve during the first 12 months of the plant's working life.

Note that the employer's remedy after the expiry of the defects liability period is a remedy in damages and not to have the works made good. This remedy under the existing law will continue in contract for 6 years from the date of completion for contracts under hand and 12 years for contracts under seal. This distinction today is totally anomalous and should be abolished. There is clear industry support for a single period of liability in both contract and the tort of negligence which it has been suggested should be for 10 years (see para. 11.9(2) of the Lathan Report).

Finally, from the purchaser's viewpoint if he wishes to be able to claim at a later date that the contract clause restricting the contractor's liability is not reasonable, then he must put forward his contentions strongly during the negotiations and retain his negotiation file papers to show that these were rejected by the contractor and he had no alternative but to contract on those terms. Even if the purchaser knows that his proposals will be rejected he should still make them, so compelling the contractor's rejection of them as a means of establishing both that the contractor was being unreasonable and that he was compelled to contract on the latter's terms.

### Dealing on standard terms of business

Where one party deals on the other's standard terms of business, that other party cannot by reference to any contract term:

- exclude or restrict any liability of his for breach of contract, or
- claim to be entitled to render a performance substantially different from that which was reasonably expected of him, or
- in respect of the whole or any part of his contractual obligations render no performance at all,

except in so far as the contract term satisfies the requirement of reasonableness (s.3 of the Act).

The primary issue here is what is meant in the construction industry by a party's 'written standard terms of business'. It is easy to see when dealing with a seller who habitually issues quotations with his standard terms printed on the back and receives orders accepting his quotation. But with large construction contracts these are usually either let on an industry standard, such as the ICE 6th, or on a contractor's or employer's home-made form from which for commercial reasons he will quite frequently depart.

It is generally not considered that dealing on an industry

standard comes within s.3, but what about the home-made form from which from time to time the contractor departs? In one of the very few cases which have been concerned with this issue, *Chester Grosvenor Hotel Co. Ltd v Alfred McAlpine Management Ltd* 1991, Judge Stannard stated that:

What is required for terms to be standard is that they should be regarded by the party who advances them as its standard terms and that it should habitually contract on those terms. If it contracts also in other terms it must be determined in any given case and as a matter of fact, whether this has occurred so frequently that the terms in question cannot be regarded as standard, and if on occasion a party has substantially modified its prepared terms, it is a question of fact whether those terms have been so altered that they must be regarded as not having been employed on that occasion.

The evidence before the judge was that within a 34-month period McAlpine had contracted seven times on the form at issue in the case, each time with some modifications, but all derived from a common base, and that over the same period they had contracted six times on their employer's form. On those facts he held that McAlpine had contracted on their standard form and therefore the relevant provisions had to pass the test of reasonableness which in fact they did.

# Performance guarantees

So far the question of defects has been considered in relation to the normal defects liability clause. But on contracts for important mechanical, electrical, or process plants there is also usually a provision that the plant must be capable of a required or guaranteed level of performance, which carries with it an acceptance by the contractor of financial liability should the terms of the guarantee not be fulfilled. If this occurs, then in order to avoid long arguments in reaching a final settlement, and to protect the contractor by fixing the total liability in advance, so that he knows this when he tenders, such a contract will normally include liquidated damages for failure in performance.

In the negotiation of such provisions the following points are important to bear in mind:

1   The guaranteed standard of performance must be clearly stated with a defined base which must be attained before the plant can be tested. Thus if the guarantee is a qualitative one, the plant must have achieved the quantitative standard before testing can start. Alternatively, one may operate the plant so as to get the quality of product required and express the guarantee in terms of through-put at that standard.

2   It must be possible to determine whether or not the guarantee is being achieved. With certain types of operation this may be difficult without at least very complex instrumentation.

3   The guarantee is normally related to a certain feed stock. If this is likely to vary, means must be established within the guarantee to adjust for this.

4   The method of testing must be clearly laid down; this is vital since different methods can easily produce different answers.

5   Details of who provides and at whose cost the labour, materials, and instrumentation for the test must be shown. The latter is particularly important since it may be very expensive. Also who is to control the plant during testing; this will normally be the contractor.

6   The procedure under the contract for deciding when the plant is ready for test, for testing and for repeating the tests if the plant has failed. Normally the costs of repeated tests are payable by the contractor.

Finally, if the plant fails its repeat tests, it is usual at that point to provide for the contractor to be released on payment of liquidated damages according to an agreed scale. While these damages are required to be calculated initially on the basis of the losses including loss of profits which it is reasonable to anticipate the employer would suffer, in practice it will usually be found that such losses are greatly in excess of what it would be reasonable to seek to impose on the contractor. They have therefore to be scaled down, and the process of scaling down and establishing a reasonable scale of damages for the contract might be taken in the following steps:

1 Fix the maximum damages which it is considered that the contractor could be asked to pay.
2 Decide on the steps of gradations in the scale – for example each 1 per cent loss in efficiency. The steps must, of course, be measurable, as substantial sums may depend upon whether the efficiency lies between one set of figures or another.
3 Establish the loss which it is reasonably estimated that the employer would suffer for each such gradation.
4 Decide on the allocation of the damages over the scale. Here the employer's and the contractor's interests are diametrically opposed. The employer wants to recover the maximum as quickly as possible, the contractor to spread it out over as long a scale as possible.

## Example
Maximum damages £50 000. Steps 1 per cent loss in efficiency.
Employer's loss £30 000 for each 1 per cent.
The employer's proposal might be £20 000 for the first 1 per cent, £15 000 for the second and third. The contractor would probably suggest a straight £10 000 per 1 per cent up to the maximum.

When asked to accept damages for loss of performance, the contractor will often ask for a bonus. It is more difficult for the buyer to justify the acceptance of a bonus in the case of performance than with time for completion. In designing the plant to satisfy the capacity or performance which he is prepared to guarantee, the contractor will usually seek to provide himself with a margin of safety, so the guarantee level is normally below the indicative design figures on which the employer has calculated his profitability. To support payment of a bonus, therefore, the contractor would have to beat the design figures indicated by him in his tender.

Another point to bear in mind when negotiating process plant guarantees is that, if the contractor is taking a licence on the process he is offering, then he will almost certainly be indemnified by his process licensor against liquidated damages up to at least 50 per cent of the process royalty he will have included in his contract price.

Reference was made, when discussing liquidated damages for delay, to the significance of the maximum. The same point can apply to liquidated damages for performance. It may be argued by the purchaser that while he is prepared to accept liquidated damages as his remedy if a plant is, say, up to 5 per cent below efficiency, after that he is free to reject the plant. Again if the purchaser wants this right, then it would be wise for him to say so expressly in his conditions of contract, since, if a damages for performance clause is included with a maximum, and no express right of rejection is reserved should the maximum be exceeded, it is doubtful whether any right of rejection would be implied, or, if it was, at what point this would be. MF/1 does now include provisions relating to performance tests (clause 35) and does refer in 35.8(c) to the purchaser having the right to reject 'where such failure of the Works would deprive the purchaser of substantially the whole of the benefit thereof'. However it is clear from the remainder of the clause that this is not the same as the results actually achieved in the tests being outside the limits of acceptability. It is presumably something worse; perhaps, say, that the plant can only be run at a loss.

The other difficulty with the clause is that it does not spell out the consequences of rejection other than to say that the purchaser can proceed in accordance with clause 49 (contractor's default). The application of that clause to a situation of rejection is however quite unclear. In the normal sense of the term, rejection would mean that the property and risk in the works reverted to the contractor and the employer would be entitled to recover all payments made together with the costs of dismantling the plant and clearing the site (comparison can be made with clause 30.5(c) of FIDIC). It does not however seem from the commentary on MF/1 that this is intended, since this refers to the possibility of the purchaser employing another contractor to complete the works which is hardly compatible with their having been rejected!

Finally, again as in the liquidated damages for delay clauses, the Unfair Contract Terms Act may come into play. It would seem that it can do so in two ways. First, under s.3(1) of the Act as discussed above, where the purchaser is dealing on the contractor's written standard terms of business. Second, the loss suffered by the purchaser may have arisen as a result of the contractor's failure to exercise reasonable care, e.g. by having made an error in his design calculations which ought to have been discovered if they had been properly checked so that his action would amount to 'negligence' within the meaning of the Act and therefore bring into operation clause 2.

The difficulty with standard terms of business is that while the standard form will usually set out the provisions covering the limitation of liability they are unlikely to state the actual sum or percentage of the contract price which constitutes the limit but leave this to be negotiated in each individual case. The position may be reached therefore that the court would decide that to impose a limit on the basis set out in the contract itself is not unreasonable but that the actual limit is. In considering whether a restriction of liability to a particular sum is reasonable or not section 24 (3) of the Act provides that regard shall be had in particular to: (a) the resources which the party seeking to rely on that term could expect to be available to him for the purpose of meeting the liability should it arise and (b) how far it was open to that party to cover himself by insurance.

'Resources' in this connection presumably means the resources available to the company as a whole and not just those being derived in profit from the contract. Also it seems likely that in the case of wholly owned subsidiaries of a group the court would take into account the financial strength of the group as a whole although how far this principle would be extended to a multinational corporation is uncertain. But how much of such resources is it reasonable for a company to be expected to put at risk on a single contract? Would the court take into account the whole trading situation of the group including their potential liabilities under other contracts? The only guidance which can be given on these issues is to be derived from the first instance decision in *St Albans City and District Council v ICL, The Times* 11

November 1994. There ICL, on a substantial contract for the supply of a computer system, had limited their liability for breach of contract to £100 000. As a result of ICL's default in the performance of their contract St Albans suffered a loss of over £1.3m in under-recovery of the community charge. In holding that the limitation of liability clause was unreasonable, the judge emphasized that:

- ICL was a very substantial company with ample resources and was a wholly owned subsidiary of STC plc, a company with record profits for the first half of 1988 of over £100m on a turnover of £1,109m
- that at the time of contract ICL had a worldwide product liability insurance cover of £50m and
- that the limit of £100 000 was small in relation to the risk and potential loss.

One point which is of obvious concern to the contractor is what happens if the court does find that the provisions are unreasonable. The Act does not appear to give the court power to amend the contract and it can only be assumed that the provision is void; rather the same as would happen if a court were to declare a purported liquidated damages clause a penalty. This would mean that the purchaser would be able to sue for damages at large if the rate of liquidated damages were considered unreasonable or if the limit were considered unreasonable presumably the contractor would continue to be liable to pay such damages at the contract rate down to whatever level of performance was actually achieved.

# 19 Insurance and indemnity

The problems relating to indemnities and insurance in respect of contracts for the purchase of plant and equipment or the carrying out of constructional work fall to be considered under two headings. First, those which arise out of defects in the plant supplied or work done, and second those which arise out of the employment on the purchaser's site of the contractor and his sub-contractors.

## Defects in plant and equipment supplied or work executed

The use by the purchaser of machinery which has been supplied in a defective condition, or which develops defects when used, may cause damage to other property of the purchaser or injury to persons – for example, the purchaser's staff employed to operate the machinery. Apart, therefore, from the costs involved in repairing both the machinery and other property of the purchaser which has suffered damage, the purchaser may find himself faced with claims for damages from persons who have

333

suffered injury. The question arises how far it is either reasonable or practical for the purchaser to seek to recover such costs or damages from the supplier or contractor.

Contracts for the purchase of large items of equipment are almost invariably governed by express terms and conditions which may have originated from either the purchaser or the contractor. Not surprisingly most purchasers, when faced with damage to their property or a claim for damages from an injured workman, which in their view is due to some defect in the plant which the manufacturer has supplied, consider that they should have a contractual remedy against the manufacturer concerned. Equally the manufacturer selling an item on which he can reasonably expect a profit of, say, £2000 is reluctant to accept a contractual risk which could involve him in the payment of damages of many times this amount, or the expenses of defending a law suit or both. His reluctance is increased by the fact that the risk is multiplied by the number of products he supplies, and must be measured, therefore, against his annual turnover.

Unless, therefore, some sensible middle course is adopted, the situation can develop into a contractual tug of war between the purchaser and contractor, which benefits nobody and wastes a great deal of time.

If it is agreed between employer and contractor that it is reasonable on a particular contract for the contractor to accept the liabilities mentioned earlier, it is suggested that negotiations could proceed on the following lines:

1   The liability of the contractor to be limited to cases where the injury or damage arises out of the contractor's negligence or breach of statutory duty, or a defect in the plant for which contractually the contractor is responsible.
2   The period of liability to be the same period as that which governs liability to make good defects in the plant itself. Once this has expired, then the contractor is under no further liability to the purchaser direct and, as regards claims from third parties, the purchaser gives to the contractor a cross-indemnity against these. However as regards personal injury

or death due to the negligence of the contractor, his liability must remain unrestricted under the Unfair Contract Terms Act in respect of contracts to which that Act applies.

3   The liability to extend only to the cost of making good the damage caused to the property or the purchaser or a third party or to meeting claims for personal injuries. Liability for loss of profits to be excluded.

4   The contractor's total liability to be limited to a sum of money for any one incident, except as regards personal injury or death as stated in 2 above. Again as regards third parties the purchaser must give to the contractor a cross-indemnity in respect of any claim made by a third party, for instance an owner of adjoining buildings for damage to his property, which is in excess of that figure. This cross-indemnity, in the same way as that under paragraph 2 above, is necessary because the property owner, not being a party to the contract, cannot be bound by its terms.

5   The contractor to be free from liability if the plant has not been properly operated and maintained in accordance with his instructions.

6   The contractor to be required to insure his liabilities. The sums which can be involved in meeting claims for injuries can be substantial, and there is no value in having an indemnity from someone.who does not possess the resources to meet the claim.

If the above is adhered to and the outcome is freely negotiated between the parties then it is considered that the contract terms would pass the test of reasonableness under the Unfair Contract Terms Act.

## Injury or damage arising out of work executed on the purchaser's site

Where the contractor is employed not merely to manufacture and deliver but also to carry out work on the employer's site in erecting, installing, or commissioning equipment, additional considerations arise and it is necessary to examine the question of indemnities and insurance in relation to the following risks:

1   Damages caused to the plant itself during erection, installation, or commissioning work.
2   Damages caused to other property of the employer.
3   Damage caused to the property of third parties.
4   The contractor's operations causing a nuisance.
5   Injuries to the contractor's own workpeople.
6   Injuries to other persons not a party to the contract.

While the above list has been set out in relation to a contract for the supply and installation of plant and equipment, the same risks and much the same principles apply in the case of contracts for the carrying out of civil engineering works or the construction of buildings.

Damage arising under numbers 1 and 2 above may be dealt with relatively simply, in that only the parties to the contract are concerned. Taking number 1 risk first, it must be appreciated that under most forms of contract the property in the plant will have passed to the employer on delivery so that, in the absence of anything to the contrary in the contract, the risk of damage to the plant will also have passed to the employer. It follows that whatever risk the employer wants to pass back to the contractor during the course of carrying out the contract must be set out specifically, and any risks not so set out are likely to be regarded (negligence apart) as remaining vested in the employer.

Clauses defining the respective obligations of the parties for such damage are normally drafted in one of two ways:

1   The contractor is made responsible for making good, at his own cost, any damage to the works which is due to the negligence of himself, his servants, agents and sub-contractors. Damage arising from any other cause must also be made good by him, but at the employer's expense.
2   Alternatively, the contractor is made responsible for making good, at his own cost, any damage to the works, howsoever caused, except to the extent that it arises from one of the

'excepted risks'. These are normally defined as the negligence of the employer and those risks which are uninsurable – for example, war, riot, etc.

The new 6th edition of the ICE conditions has corrected the anomaly noted in the previous edition of this book that under clause 20(3) the contractor could be liable for damage even if caused by the employer's negligence. While the contractor would still be obliged to make good the damage if it were caused by the employer's negligence, he would have a claim against the employer for an indemnity.

There are two very important distinctions between these two clauses:

1   Under the first clause it is up to the employer to show that the contractor has been negligent. As the lawyers would put it 'the burden of proof rests on the employer'. Under the second clause it is the other way round, and it is the contractor who has to show that the employer has been negligent in order to escape liability.
2   If the damage or loss is truly accidental and cannot be shown to be due to the negligence of either party, then under the first clause it is the employer who pays, whilst under the second clause it is the contractor.

Much the same applies to the second risk listed above – that of damage to property of the employer other than the works. The employer is obviously concerned that he is not called upon to pay for making good damage to his own property which has arisen out of the carrying out of the contract work. At the same time, to prove negligence can be difficult and cost- and time-consuming in itself, so from the employer's point of view it is suggested that here again he should seek to make the contractor liable for making good any such damage, howsoever it occurs, unless it is due to one of the 'expected risks'.

The remaining risks referred to above are all cases where the employer's interest is to ensure that he is not called upon to pay damages to a third party arising out of the execution of the contract.

It might be expected that, where an employer places a contract with a contractor, the liability for any accident or injury arising out of the execution of that contract, in the absence of anything in the contract to the contrary, would rest with the contractor. This is not, however, always the case and the present position under English law may be summarized as follows:

1   The duty of the employer towards third parties to the contract may fall into one of two categories:

   • a duty to take reasonable care himself
   • a duty to see that care is taken by others for whom he has a responsibility

2   Generally (other than in cases of nuisance as to which see paragraph 3 below), an employer is not liable for the acts or default of an independent contractor provided that he has appointed an apparently competent contractor to undertake the work. The more difficult question is what duty the employer has, if any, to supervise the work of the independent contractor.

   If the case falls under the Occupiers Liability Act 1957, the employer (as occupier) must be able to show that 'he has taken such steps (if any) as he reasonably ought in order to satisfy himself that the contractor was competent and the work had been properly done', and this latter expression has been said in the House of Lords to include work in progress (per Lord Keith in *Ferguson v Welsh & Others* [1987] 3 AER 777). In that case it was held that there was no general obligation to supervise but that if the employer suspects that an unsafe method of working is being used then it may be his duty to instruct the contractor to change the method to one which is safe. That case concerned the demolition of a building and it is interesting to note that 'demolition' work was held to be within the scope of s.2(4)(b) of the Act. However, it is considered that in a case of large constructional works the employer (occupier) would probably only satisfy his obligations under the Act if he had appointed professional advisers to supervise the work on his behalf.

As regards cases not falling under the Act, then towards persons with whom he has a relationship of proximity sufficiently close to establish a duty of care, the employer may in particular circumstances owe a duty to take reasonable care to supervise the work of an independent contractor. Thus property developers entering into a contract for the construction of factory premises and entrusting the work to an associate company were held liable in negligence to the person for whom the factory was being constructed for their total failure to do anything to supervise the construction work (*Cynat Products Ltd v Landbuild (Investment and Property) Ltd* [1984] 3 AER 513). However, although the practice is now widespread in the construction industry of employing small sub-contractors, or sub-sub-contractors often on a labour only basis who have few assets, in general there is no legal restriction on the main contractor from doing so and no liability will be attached to the main contractor towards a third party in negligence if the sub-contractor is in breach of his statutory or common-law duties. Only exceptionally, if the main contractor was aware that the sub-contractor was performing his work defectively and in a foreseeably dangerous way, could the main contractor be potentially liable as a joint tortfeasor with the sub-contractor (*D & F Estates v Church Commissioners for England* in the House of Lords [1988] 3 WLR 368, reversing the decision at first instance to which reference was made in the previous edition of this book).

It must be stressed, however, that in all these cases if the employer is liable it is only because he has broken his primary duty of care towards the injured party and it must first be established that he owes such a duty of care. He is not vicariously liable for the negligence of the independent contractor in the way in which he would be for one of his employees.

3   If the duty is to see that care is taken then the employer

cannot delegate his responsibilities by employing an independent contractor. The most common situation in which this duty occurs is when an absolute obligation is placed upon the employer by statute, e.g. the obligation to fence dangerous machinery under the Factories Act. Another rather less common situation is that of nuisance which is essentially an act or omission by which an occupier of land is disturbed in his enjoyment of it. This can take the form of actual damage to the land but is more often a matter of causing a loss of comfort or convenience, for example through the escape of smells, noise or dirt, and is usually the result of activities of a neighbour or those for whom he is responsible which may include independent contractors. The distinguishing feature of nuisance is that once the facts of the nuisance have been established, i.e. the level of damage or discomfort is not one which the person affected should be expected to put up with, then it is no defence that all possible skill and care have been taken to prevent it. 'If the operation cannot by any skill or care be prevented from causing harm then it cannot lawfully be undertaken at all except by the consent of those affected or by the authority of statute.' Proof is concerned, therefore, to ensure that, where any damage or injury arises out of the contractor's default, it is the contractor and not the employer who has to meet the claim by the third party. Accordingly the employer's first step is to obtain an indemnity in the contract from the contractor under which the contractor undertakes to indemnify the employer against any claims made against the employer by third parties and any costs, damages or expenses which the employer may be called upon to pay.

In the drafting of the provisions relating to indemnity the same point arises again as has been discussed above. Is the contractor to be liable only for the consequences of his negligence or breach of statutory duty, or does his liability extend to cover any claim for damage or injury arising out of the carrying out of the contract work unless this is due to one of certain specified 'excepted risks'?

It is interesting to note that the current editions of three forms of contract prepared and issued by professional associations each solve this problem in a different way, as follows:

## 1 Institution of Civil Engineers (ICE conditions of contract)

The contractor is liable to indemnify the employer against any damage to the property of third parties or injuries to persons arising out of the execution of the contract, except to the extent that such damage or injury is caused by the negligence of the employer or is due to one of certain other stated 'excepted risks'. The burden is therefore placed wholly on the contractor unless he can bring himself within the exception provisions, and the onus of proof is on him to do so.

## 2 Joint Contracts Tribunal (JCT form of contract)

The conditions provide that the contractor is absolutely liable for injury to persons except where he can show that such injury arose from an act or default of the employer. In the case of damage to property, however, the emphasis is the other way round, and the contractor is only required to indemnify the employer where it can be shown that the damage arose out of the contractor's negligence.

## 3 MF/1

The liability of the contractor has been extensively modified in comparison with that which prevailed under the old Model Form A. In essence before take over the contractor indemnifies the purchaser against any third party claims which arise out of the execution of the works unless these are due to the negligence of the purchaser. This liability does not require proof of negligence. After take over the contractor gives the purchaser an indemnity against such claims to the extent they are due to his negligence.

It must be stressed that the fact of the employer having obtained an indemnity from the contractor does not in any way lessen the employer's own legal liability, and the third party is perfectly free, if he can establish a valid claim, to proceed against the employer. It is of the utmost importance, therefore, to the employer that the contractor has adequate resources available to implement the terms of the indemnity, and it is suggested, therefore, that having as a first step obtained an indemnity, the employer must, for his own protection, take two further steps:

- require the contractor to take out the necessary insurances
- check that the contractor has in fact done so and that the policies concerned properly cover the risks against which the contractor is required under the contract to give the indemnity.

As to the first, the contract should require specifically that the contractor does take out insurance against all the risks which he is assuming and not, as is the case in certain standard forms, merely against damage to the works through fire. It is normal, when requiring a contractor to take out an insurance against third party claims, to indicate the minimum value for which the policy is to be taken out. Most forms of contract which do this state that such minimum value is for insurance purposes only and does not represent a limit of liability. It is clearly correct to do this, particularly as with large companies carrying 'blanket' insurance policies the limit of liability specified in the contract may well be much lower than that included in the policy. It is, however, as well to recognize that in practice the limit for which the contractor has insured is the most which the employer is likely to recover, at least without putting the contractor into liquidation.

The second problem is more difficult. How is the employer to be sure that the contractor has carried out his contractual duties properly and that there are not exceptions or exclusions within the policy which are inconsistent with the contractor's obligations?

Most of the professional institute forms of contract provide for the contractor to produce his policies and the current premium receipts therefore to the employer for his inspection. It is doubted, first, how far this is carried out in practice and, second, whether the employer gains any great benefit from such inspection in those cases where it is done. Insurance policies are technical documents, often of considerable complexity, and it requires an expert in insurance to check that the policy is in fact in conformity with the terms of the contract. It could, moreover, be most embarrassing if subsequent events were to show that the policy did not in fact fully cover the contractor's liabilities. There is a further practical objection to the inspection of policies, which is that many companies have 'blanket' policies and do not insure each contract separately. Obviously in such circumstances they would not wish such a policy to be continually sent for inspection to various clients.

Bearing in mind that what the employer really wants to know is simply: (*a*) the contractor does have a policy covering the contract, and (*b*) that there are no endorsements or qualifications on the policy which affect the risks involved on the contract, it is suggested that the contractor should merely be required to supply a certificate to the employer from his insurance company or broker to the effect that the contractor is insured against the risks detailed in the contract and listing any exclusions or qualifications to the insurance cover. It is recognized that a system of this sort is not foolproof, and that if the broker or insurance company made a mistake in the certificate, then his only remedy would be a claim in negligence under the principles established in *Hedley Byrne & Co. v Heller & Partners Ltd*. On the other hand it has the merit of simplicity, it can be operated in general by staff not possessing expert knowledge on insurance matters, and it is, accordingly, that much more likely in fact to be carried out in practice. This is the vital factor. It is not the slightest use having a foolproof system if nobody operates it.

## Owner-controlled or project insurance

As an alternative to the insurance provisions under the standard forms of contract it is worth the employer considering on larger projects taking out insurance cover himself, both in respect of the works and of public liability for the benefit of all engaged in the project. The insurance would cover therefore the contractors, sub-contractors, suppliers and consultants on a non-recourse basis, but with significant deductibles and with certain limited risks, e.g. for motor vehicles left with the contractor.

This gives the employer control of the insurance position, knowing that proper cover has been taken out and maintained. For this reason it is usually required where the project is being financed on a project finance basis. It may also reduce the overall insurance costs.

### Latent defects insurance

In relation to the contract works there are two main deficiencies in the current insurance provisions. First, in general the contractor's insurance will not cover the cost of making good the actual defects themselves. Second, once the works have been taken over the contractor's insurance will only cover for damage which he causes carrying out remedial work or which arises from a cause originating prior to the commencement of the defects liability period.

It has therefore been proposed – see the Build Report published by NEDO in October 1988 – that there should be latent defects insurance cover for building foundations and structures which would run for a period of 10 years from practical completion. It would be with the waiver of subrogation rights against all those involved in producing the building, but with realistic deductibles to ensure discipline.

This proposal was supported in the Latham Report with the recommendation that such insurance should become compulsory for all future new retail and commercial buildings (see Chapter 11 and recommendation 11.24).

The same problems relating to the need for indemnities and insurance as referred to above arise on overseas contracts but two additional difficulties may be present. First, as regards any cross-indemnity obtained from the purchaser, its value in practical terms will depend on its enforceability in the territory in question – often a matter of considerable doubt. Despite therefore having obtained the indemnity the contractor may need to consider arranging his insurance cover as if no such cross-indemnity had been given.

Second, if the purchaser is a foreign government or quasi-governmental agency, it is likely that they may require the contractor to insure his risks through a national insurance company, if one exists, or if not through one which is a locally owned company. Such a company may be rather more adept at receiving premiums than paying out claims (as was the case with Bimeh Iran in the days of the Shah) and will also only pay out claims when it does do so in local currency which may either be non-convertible or at least subject to exchange control. Assuming the requirement to insure with the national company is a statutory one there are two possible solutions to the problem. The first is to take out additional insurance in the form of a difference in conditions policy with a UK company under which the UK company will pay the claim in the UK and the contractor pays the company any moneys which it does manage to recover from the national company of the territory in question. Beware of a policy written the other way round under which the UK company only makes up the non-recovered balance since this may take years to establish. With this solution there can be a practical problem in obtaining access for the UK company's loss assessors to inspect the damage and certain subterfuges may need to be adopted.

Second, since the national insurance company or one locally owned will almost certainly have reinsured all but a small proportion of the risk either in the UK or Europe it may be possible to obtain a 'cut-through' agreement with the lead reinsurance underwriter so that the contractor can deal with him direct. This is more likely to be practical if the requirement is to insure with a local company which is itself perhaps partially owned or associated with a UK or European insurer.

One final point: as in other matters, whatever risk the employer transfers from himself to the contractor has a price attached to it, and the employer is going to be called upon to pay that price. It is sometimes suggested that all problems can be solved by making the contractor liable, but this is nonsense. If there are special risks involved in the execution of particular work, then this sort of problem can only be solved by cooperation between employer and contractor with each assuming his fair share of the risks involved, rather than by simply seeking to pass the problem over to the contractor. For this reason the employer on any large project should certainly consider the alternative, as regards the insurance of the works, of there being one policy, taken out by himself, and expressed to be in the joint names of the employer, the main contractor and all sub-contractors and sub-suppliers.

# 20 Fixed price and price escalation

There is still a continuing debate on both sides of industry as to the respective merits and demerits of contracts on a fixed price as opposed to those subject to escalation. Although the attempt is sometimes made from the purchaser's side to relate this debate to the broader issue of combating inflation it is considered that this is ill-conceived. The purchaser's task is to procure the goods and services he requires on the most favourable overall economic terms relative to the strength of his negotiating position. If for example that position is weak then it would be an act of commercial folly for him to attempt to negotiate a fixed price, because that was regarded as contributing to the fight against inflation, when the element of 'cover' which the contractor would be able to insist on was included within his price and would almost certainly exceed the amount which would be recoverable under a properly constructed escalation formula.

Factors which it is considered are relevant to the choice of which method to adopt are as follows.

### Fixed price – advantages to the contractor

1   The amount for escalation is established within the price, so avoiding any post-contract administration problems or debate on the interpretation of escalation provisions.
2   Since the amount of the 'cover' is spread across the prices the contractor may be able to improve his cash flow and the cover will be subject to profit.
3   The contractor is not required to make any disclosure of the build-up of his prices.

### Fixed price – disadvantages to the contractor

1   It may be very difficult to estimate the amount of the 'cover' required so that the contractor becomes exposed to serious loss if he underestimates. This was the position in which many firms found themselves when the price of oil dramatically escalated as did that of oil-based chemicals. The extent to which this is so, and the significance of the 'cover' in relation to the contract price, will be largely a function of the time period of the contract, the general anticipated level of the applicable inflation rates and the degree to which the contract work includes any element which is subject to particularly unpredictable price changes. One such for many years has been copper.
2   The severity of the competition may prevent the contractor, if he wishes his bid to have a high success probability, from including an adequate element of cover. By contrast a price escalation clause removes that element from the competition.

The worst combination of circumstances from the contractor's viewpoint is strong competition, with too many firms chasing too little work, and high inflation. It might have been thought at one time that this combination either could not occur, or at least would be very temporary, since the effect of competition would be to cause costs and therefore inflation to fall. However, market situations for both labour and raw materials have shown in recent years strong resistance to downward pressures and it is this very combination of circumstances which contractors, particularly in the international business, are faced with today.

## Fixed price – advantages to the purchaser

1   The purchaser knows his total commitment in advance instead of having the possibility of unpredictable cost overruns.
2   Where competition is present the 'cover' included by the contractor is likely to be less than the amount which is recovered under the escalation clause.
3   The contractor has a greater incentive to seek to control his costs both for materials and labour.

## Fixed price – disadvantages to the purchaser

1   In the absence of competition, or the less effective the competition, the more opportunity exists for contractors to load their prices by inflated allowances for escalation.
2   If the contract work has not been totally defined and extensive variations become necessary, the difficulty of their negotiation is increased. The contractor is bound to tie his fixed price to the scope of work and programme as originally established.
3   If the contractor's cover, either through his inability to assess future events correctly, or the severity of the competition, is grossly inadequate, the purchaser may be landed with a contractor who is losing money and whose contract performance is likely to suffer accordingly.

Summarizing the above from the viewpoint of both sides, if the risk element in quoting a fixed price can be reasonably well defined, does not constitute too significant a proportion of the total contract price, and if there is effective competition, then both sides would be likely to prefer a fixed price. If any of these three assumptions are not correct the balance of advantage for both sides would seem to lie in favour of at least limited escalation. That is, if the problem can be restricted to particular materials, or even to wage rates in particular territories, these could be made subject to escalation while the rest of the contract is placed on a fixed price.

So far the problem has been looked at on the basis of a single relatively short-term contract in which the degree of control which a contractor can exercise over increases in labour costs is limited.

On the assumption that he is employing union labour, he is bound to apply awards made under national agreements. Further, he can often only seek to contain, rather than totally reject, the wage creep which follows or supplements such agreements. He has no control over indirect charges related to labour, such as national insurance.

What a contractor or supplier can do is to try to offset the increase in rates or charges by an increase in productivity and a decrease in the labour force employed, so that the total increase in cost to the company for labour is minimized. Such measures are often, however, essentially long-term operations. For success they may require a change in working methods or the introduction of new plant or machinery. Certainly the employers will need union co-operation, which is bound to take time to obtain. There is likely, therefore, to be little immediate opportunity for achieving counter-balancing savings in this way on short-term contracts.

It also follows that a contractor's current price for doing a particular job after application of the total effect of a wage increase may be higher than the price he would tender for the same work in six or twelve months' time, when he will have had a chance to apply cost reductions to offset the additions in wages. This is a point which has to be borne very much in mind when dealing with long-term repair or other forms of long-term contracts which are labour intensive. Continued application of straight wage increases over a substantial period of time is almost certain to inflate the contract price and provide the contractor with extra profit.

There must accordingly be a difference in approach between one-off contracts of a limited duration for specific work and long-term purchasing agreements or running contracts. With the former it is reasonable to ignore the productivity or cost offset factor; with the latter any negotiations must take this into account either immediately or by providing for a price review in, say, six months' time. This could be coupled with a target cost reduction which the manufacturer is asked to achieve and some sharing arrangement so that he has a profit incentive to do so.

## Methods of determining price escalation

There are two alternative methods which may be used in order to calculate the amount either due to the contractor for increases in costs or to be deducted from the contract price for reductions.

The first is usually referred to as the 'actual cost' method and is the one which has traditionally been used in the building and civil engineering industry. With this method the determination of the increase/decrease is by way of comparison between a schedule of labour rates and material prices included in the tender and the increase/decrease in such rates and prices allowed for under the price variation clause multiplied by the quantities of labour and material utilized on the contract after the date of the increase/decrease. Reimbursement of material fluctuations is usually restricted to basic list and there is normally no specific reimbursement for profit and overheads. There is no particular problem with the mechanics of this method for building and civil engineering contracts; only the tedium for the contractor's staff of preparing the claims and of the quantity surveyor's staff of checking and agreeing these.

With manufactured goods the matter is not quite so simple. The contractor's tender, which forms the basis from which the comparison must be made, will not normally contain any detailed information as to the labour rates and material prices on which the tender is based. Moreover, the prices for bought-in items will often be assessed from estimated data rather than derived from firm quotations. This is inevitable if the time required for and the costs of tendering are to be kept within reason. It follows that, to make the comparison, it is necessary to go behind the tender to the detail of the contractor's estimate, including the estimates of manufacturing costs. Manufacturers are often reluctant to disclose such details and suggest as an alternative to doing so that the amount of increases should be certified by their auditors. If the matter is treated purely as one of arithmetic then this is not an unreasonable solution to the problem. It saves both manufacturer and purchaser a great deal of time and effort, and the purchaser has the assurance that an independent body has examined and certified the costs.

If, however, the issue is to be looked at not solely as one of arithmetic; if the purchaser is to ask the question not only 'Did the increase occur?' but 'Should it have occurred?' then it becomes an entirely different matter. Take for example manufactured goods which include a component which is available from five suppliers. When tendering, the manufacturer rings up supplier *A* and is quoted over the telephone a price of £250, which he includes in his estimate. When he comes to place the order *A* submits a written quotation for £275, which the manufacturer accepts, and then claims the £25 under the price variation clause in the contract. On the face of it he is justified in doing so, and it is believed that an auditor would accept it. But supposing the purchaser could show that if the manufacturer had invited competitive quotations from the other four suppliers he could have obtained a quote from *D* of £245?

This is a simple case; others are more difficult. There may be variations in quantity between those on which the estimate was prepared and those on which the job is built due to development of design. The specification for a bought-in item may change, so that a new quotation has to be obtained which cannot strictly be compared with that on which the estimate was prepared.

If the factor of efficiency is to be taken into account, if the purchaser is determined to ensure that estimating errors or deliberate under-pricing are not corrected through the medium of the price variation clause, then the determination of the amount properly due to the manufacturer becomes a major exercise. Moreover, it is an exercise which by its nature cannot be carried out until the work is complete, by which time it is often difficult to ascertain just what did happen, and why, some two years earlier.

For these reasons it is common, particularly with manufacturing contracts, for price variation to be determined by means of a formula under which the increases or decreases are arrived at by reference not to a fixed individual cost, but to independent indices which reflect variations in national labour and material costs. Obviously a formula constructed in this way is never likely to give the right answer for any one contract, but it should come near enough to be acceptable when averaged over a number of contracts.

This point is important as it is sometimes maintained that if it can be shown that the formula will result in an over-recovery of the contractor's costs then the amount to be paid to the contractor should be reduced. That is not however so. By having selected a formula as the method by which price escalation is to be determined the parties will be taken to have agreed to be bound by the outcome whichever way it goes.

The data required for the construction of a formula are as follows:

1   The percentage of the contract price which it is to be assumed for the purpose of the formula is fixed. In a period of continuing high inflation coupled with high interest rates it can be argued that in reality nothing is unaffected by inflation. Even profit, if it is to reflect the depreciation in the value of money over the contract period and give a satisfactory level of return on the inflated costs of performing the contract, should be treated as a variable. The most widely used formula in manufacturing is that of the British Electrical and Allied Manufacturers' Association (the BEAMA formula) which now suggests only a 5 per cent fixed element and many contracts are in fact placed with 100 per cent of the price variable.

2   The proportion of the contract price representing labour and materials. These will obviously vary according to the nature of the contract work being undertaken. The standard BEAMA formula assumes that for electrical equipment and general mechanical engineering work they are equal. There is an obvious advantage to the purchaser in an inflationary period in seeking to reduce the proportion related to labour since the changes in labour rates are reflected immediately in the index and the rate of change in the labour index is generally greater than that for materials.

Although for simplicity it is preferable to have one set of

proportions applied to the whole contract, if there is a significant and identifiable portion for which the ratio is markedly different from the rest, then it may be necessary to apply separate ratios to each.

3   The definition of the contract period. The BEAMA formula proposes that for home contracts excluding erection this should be the period between the date of order and the date when any portion of the plant which is invoiced separately is ready for despatch. When the contract includes erection then the period should be between the date of order and the date when the plant is taken over or ready for commercial use.

For contracts excluding erection the definition of the contract period is reasonable. However with many contracts for supply and erection manufacture will be finished some time before the end of the contract period if this is taken to be when the plant is completed ready for commercial operation. During this final stage only erection and commissioning will be involved and if it is of considerable duration then the effect of including it within the contract period will be to distort the operation of the formula. There is a proviso to the BEAMA formula which allows a shorter period corresponding to the manufacturing cycle of the plant to be adopted if this is stipulated in the tender or agreed in the contract. It is important for the purchasing officer to be aware of this and to require in his invitation to tender for the tenderers to submit their bids on this basis.

Alternatively the contract may be divided into two parts: one for manufacture, as if the contract did not include erection; and the other for erection only with separate formulae applying to each, and this it is suggested is preferred.

4   The portion of the contract period over which it is assumed that manufacture takes place. Manufacture and therefore the incidence of labour costs does not commence until an appreciable part of the contract period has expired. The broad assumption normally made is that expenditure on labour is not significant until about the one-third stage of the contract period and that thereafter it continues at a uniform

rate. In practice one would expect a tailing off of expenditure during the last stages of manufacture when final assembly and testing take place, but to try to take this type of factor into account would undermine the simplicity of the formula which is its greatest attraction.

5 The portion of the contract period over which it can reasonably be assumed that materials are purchased. With materials it is obvious that their purchase starts early on but is completed before the end of the contract period. Again for the sake of simplicity it is convenient to end the materials period at what can be assumed to be the mean point when expenditure on materials purchasing is incurred and this is generally judged to be three-fifths of the contract period.

However BEAMA have in the interests of accuracy decided that because of the irregular movement of the indices of materials it would be more appropriate to use the average of the indices between two-fifths and four-fifths of the contract period rather than a single date at the end of the three-fifths period. While these periods are generally applicable there is no particular magic about them and if there is a particular contract for which their use would appear inappropriate then it is up to the purchasing officer to propose in his inquiry that different periods should be used.

6 An independent index which it is reasonable to apply as a measure of wage increases or decreases. It is important to note the extent to which any such index includes not merely wage increases but also makes allowance for changes in statutory charges such as national insurance contributions. The BEAMA labour cost index is based upon the national average earnings index figure for the engineering industry produced by the Department of Employment and takes into account national insurance statutory payments, nationally agreed variations in paid holidays and the working week. Two figures are produced: one for the electrical industry and one for the mechanical.

7 An index which can reasonably be applied to measure increases or decreases in materials costs. More than one index may be necessary if, for example, both copper and steel are

used extensively in manufacturing the plant since these move independently of each other. However the temptation should be resisted of complicating the formula by including too many indices related to only small proportions of the materials cost.

BEAMA publish a number of these indices obtained from the Department of Trade and Industry and the appropriate choice should be made for the contract in question of which of the DTI indices are the most relevant.

8    The base labour and material indices which the purchasing officer should require the contractor to declare in his tender.

From the above data, in order to arrive at an assessment of the variation in costs since the date of tender, it is necessary to compare: the base labour index with the average of the indices for the last two-thirds of the contract period; and the base materials index or indices with the average of those between the two-fifths and four-fifths of the contract period.

Mathematically the formula can be written as:

$$P = Po\left[ F + \left( Le\frac{L}{Lo} \right) + \left( Me\frac{M}{Mo} \right) \right] - Po$$

*where* $P$ = the price increase or decrease

$Po$ = the original contract price

$F$ = the fixed element, if any

$Le$ = the proportion of the contract price variable with labour usually referred to as 'the labour element'

$L$ = the current average labour index

$Lo$ = the labour index at the base date of the tender

$Me$ = the proportion of the contract price variable with materials, usually referred to as 'the materials element'

$M$ = the average materials index over the materials period

$Mo$ = the material index at the base date of the tender

Although originally the BEAMA formula was intended to be

applied at the end of the contract period it is now common particularly on export contracts for the formula to be applied on applications for monthly certificates for progress payments in respect of the value of work completed. In this event the contract period is taken as the period between the date of order and the date of application for the certificate concerned. There is then deducted from the amount of escalation so calculated the amount of escalation paid on the previous certificates.

It can be argued that this method provides the contractor in a period of inflation with an over-recovery in that the higher escalation rate is applied to the gross value of work certified to date and yet in so far as it is work which has been previously certified and paid for the work must have been completed and so should no longer attract escalation. An alternative method would be to apply the new rate calculated each month only to the movement in the monthly valuation.

An example to set out the difference is given below:

| MONTH | VALUE CERTIFIED | PER CENT INCREASE | ESCALATION PAYABLE | | |
|---|---|---|---|---|---|
| | | | | Method 1 | Method 2 |
| 1 | 100 000 | 10 | | 10 000 | 10 000 |
| 2 | 120 000 | 11 | 13 200–10 000 | 3 200 | 2 200 |
| 3 | 150 000 | 12 | 18 000–13 200 | 4 800 | 3 600 |
| | | | | 18 000 | 15 800 |

It is recommended that if it is intended to pay escalation calculated on monthly interim certificates then the method of calculation should be carefully considered and set out in the invitation to tender documents.

Three other difficulties of interpretation and application may arise. The first is the definition of the order date. It is not unusual on certain large contracts for a letter of intent to be issued in advance of the placing of the formal order or the signature of the contract, so that work can proceed to the financial limit stated in the letter of intent. In this event the letter of intent should make it clear that the order date for Contract Price Adjustment (CPA) purposes should be the letter of intent, at least to the value of the financial authority given, and if this is subsequently increased, then the same procedure should be applied. If this is not done, and the formal order or contract signature is delayed for any significant time, the contractor may well argue that the order date for CPA purposes, which starts the contract period running, is the date of the formal order or contract and this could make a great difference to the amount of escalation payable. It would seem that the contractor's argument is probably contractually correct and even though it would result in an over-recovery the purchaser would be obliged to pay. Having selected the formula as the means by which the escalation is to be calculated it is not open to the purchaser afterwards to maintain that it gives the wrong result.

The second is the definition of the date of delivery. The BEAMA formula provides for the situation in which a large contract is completed in sections but not for where a section is delivered over a period of time for site assembly. It would clearly be unreasonable from the purchaser's viewpoint to allow delivery to be defined as when the whole unit had been delivered as this could mean that a greater proportion by value of the unit was still accumulating escalation although it had been for many months on the purchaser's site.It is suggested that for each main unit there should be a definition of what constitutes delivery by reference to the delivery being deemed to have taken place either when an identifiable part has been delivered or when parts to a certain value have been delivered.

The third problem area is the one which arose in the case already referred to (see p. 200) of *GLC v Cleveland Bridge and Engineering* which concerns the application of the proviso to the BEAMA formula that 'no account shall be taken of any amount by which any cost incurred by the contractor has been increased by the default or negligence of the contractor'. It was held in *Cleveland Bridge* that if there is no further contractual obligation other than to complete by a certain date, then it is for the contractor to plan and perform his work as he wishes. If that results in certain items being manufactured later than they could have been as a result of which the employer is called upon to pay a greater sum by way of escalation than he would have been had they been manufactured earlier. Then that is not default or negligence on the part of the contractor. It follows therefore that on a project with a long completion period if the employer wishes to protect himself against the risk of additional escalation due to manufacture being delayed to suit the contractor's convenience, then he must make it a contractual obligation for the contractor to comply with some specific manufacturing programme. Of course that carries with it the other implications, as the Court of Appeal pointed out, that the employer would then have to pay for such items and take delivery of them earlier (or pay storage charges) and also the employer must be ready to provide the contractor with any drawings, data or other services so that the contractor is enabled to meet that programme.

# Installation

The erection element of a plant contract should be treated separately for the purposes of the escalation calculation. On UK contracts it would generally be sufficient to provide that the erection price is adjusted by applying to the value of work executed each month the difference between the base labour rate and the labour rate applicable to the month in question. That is, the erection portion of the contract price is treated as wholly variable with labour, and the fact that the price will include for certain element of cost not strictly wholly labour-related, e.g. site huttage, cranage etc. is ignored as not being of sufficient significance to justify complicating the formula.

However on overseas contracts there are different considerations. First it is likely that a greater proportion of the installation cost will comprise either locally recruited or third national labour for which it would be inappropriate to use a UK labour index. Second, the costs of the UK staff will include housing, transportation, messing etc. which again are related to local conditions.

At the very least therefore the price should be divided into two elements: UK salaries and salary-related benefits which can properly be adjusted by reference to a UK labour index; and the local element which is adjusted by reference either to a local cost of living index or labour index. The difficulty often encountered is to discover a suitable local index which is both reliable and regularly published. If a main element is third national labour then this will need separate treatment as its wages and benefits are usually tied to the regulations of its own country.

If no suitable index exists then it may be necessary to construct one using as a basis the rates actually paid for a spread of trades and the proportion which each such trade represents of the total labour cost. This proportion is then assumed to be static over the contract period. The expression of the formula then becomes:

$$PF = W_1 \left[ \frac{L_0 - L^1}{L^1} \right] + W_2 \left[ \frac{L_0^2 - L_0}{L^2} \right]$$

*where PF* = the price variation factor
   *W* = the weighting factor for the particular trade
   $L_0$ = the labour rate for the month in question
   *L* = the base labour rate

In practice the calculation would be done at probably no lesser intervals than three months. Note that it is essential that the calculation is done in this manner and not by adding the sum of the labour rates times the weighting at the base date and comparing it with an equivalent sum for the month in question. A simple example will show the difference between the correct and incorrect methods.

Assume two classes of labour: tradesmen and labourers. The relevant data are as follows:

   rate for tradesmen at base date 100; at the month in question 120

   rate for labourers at base date 50; at the month in question 55 weighting for tradesmen 0.7; for labourers 0.3.

## Correct method

$$0.7 \times \left(\frac{120-100}{100}\right) + 0.3\left(\frac{55-50}{50}\right) = \left(0.7 \times \frac{20}{100}\right) + \left(0.3 \times \frac{5}{50}\right)$$
$$= \frac{14}{100} + \frac{3}{100}$$
$$= 17\%$$

## Incorrect method

$$100 \times 0.7 = 70 \qquad 120 \times 0.7 = 84$$
$$50 \times 0.3 = \frac{15}{85} \qquad 55 \times 0.3 = \frac{16.5}{100.5}$$
$$\% \text{ increase} = \frac{100.5 - 85}{85} = \frac{15.5}{85} = 18.2\%$$

It can readily be seen that if the labourers had received no increase at all the first method would have given an increase of 17 per cent which is obviously correct while the second method would have given an increase of approximately 18.2 per cent which is obviously wrong. The importance of using the correct method is emphasized by the practical consideration that in territories which do not have readily available reliable indices it is usual for the skilled workers to receive much larger increases than the labourers.

# Building and civil engineering

In recent years price escalation formulae have been introduced both for building and civil engineering in the UK and their principles applied to work overseas. The two formulae, the Baxter formula for civil engineering works and the Osborne formula for building work differ somewhat in the method of their operation and only a generalized description of each is within the scope of this work. For more detailed explanations of their operation and interpretation reference should be made to the standard guides to the formulae published by HMSO. One general comment on both formulae, both much more so with the Osborne formula, is that in the eagerness to produce formulae which give a genuine representation of the fluctuations in cost incurred by contractors on particular contracts, those concerned have in the author's view produced formulae which are too complicated and cumbersome. It has to be recognized that price variation formulae are only approximations so that there will be swings and roundabouts. Sometimes the contractor will gain and sometimes the employer. Unfortunately this type of approach does not appear to be readily accepted by the quantity surveying profession. To justify its use a price variation formula must be simple and cheap both to set up and administer and it is doubted whether the Osborne formula can be so described.

### The Baxter formula

The contract price is sub-divided into three elements: labour, materials and plant; and the materials portion is further subdivided into a number of items such as aggregates, cement, reinforcement etc. There is excluded from the contract price the value of nominated sub-contractors and dayworks. Each of these elements as a percentage of the contract price is then multiplied by the percentage which is to be treated as fixed to arrive at the value of the coefficients for the purpose of the application of the formula. A simplified example is given in Table 20.1.

The effective value of each certificate issued by the engineer monthly is then adjusted by a percentage which is the sum of multiplying each coefficient by the current index minus the basic index divided by the basic index. Using the above figures then as an example, if the basic index for labour was 100 and the current index 120 the labour adjustment would be:

**Table 20.1 Example of the Baxter formula**

| ITEM | BREAKDOWN OF CONTRACT VALUE SUBJECT TO PRICE ADJUSTMENT % | VALUE OF COEFFICIENT, COLUMN 2 × 0.9 |
|---|---|---|
| Labour | 35 | 0.315 |
| Plant | 20 | 0.18 |
| Materials: | | |
|   aggregates | 15 | 0.135 |
|   cement | 6 | 0.054 |
|   fuel | 8 | 0.072 |
|   reinforcement etc. | 16 | 0.144 |
| Non-adjustable element | | 0.10 |
| | 100 | 1.00 |

$$0.2975 \times \frac{120 - 100}{100}$$

Each coefficient is treated in the same manner, the resultant sum is then multiplied by the effective value of the certificate to give the cost adjustment. In this formula:

The effective value is the amount which is due to the contractor less any amounts for dayworks or nominated sub-contractors or other sums which are payable on actual cost and any sums for increased or decreased costs under the formula less the amount previously certified again with the same exclusions.

The basic index is the index applicable two days before the return of tenders.

The current index is the index appropriate to the certificate in question. The indices are prepared by the Department of the Environment and issued by HMSO.

It should be particularly noted that the subdivision of the contract price into labour, plant and materials and the calculation of the coefficients for inclusion in the formula is carried out only once when preparing the contract documents.

No attempt is made to adjust these for the individual certificates to give effect to the actual proportions which relate to the work included within that certificate. This is rough justice but it is a triumph for simplicity and indeed once the percentages are established the actual operation of the formula should present no difficulty.

Structural steelwork if it represents a substantial proportion of the contract price is treated under a separate formula.

If the contract completion is delayed and no extension of time is granted, the value of the indices ruling six weeks before the completion date are used for all certificates issued after the completion date so the contractor gains no benefit out of his own default.

## The Osborne formula

The basic concept of the formula is that the contract work to be performed by the main contractor, e.g. exclusive of nominated sub-contractors against prime cost sums, is divided into a number of work categories, the total list is 38, related to the work sections of the standard method of measurement, e.g. formwork, brickwork, softwood joinery etc. Each work category has its own index which is a compound of indices for labour, materials and plant and these are published monthly by the Department of the Environment.

Each item in the bill of quantities is annotated with the number of the work category into which it falls, although necessarily some items for preliminaries will not have a work category number and fall into what is described as 'the balance of adjustable work'.

On each valuation the total value of work executed for each work category priced at tender rates is established and then multiplied by the current index for that category minus the base index divided by the base index. The result is then summed to give the total fluctuation payable for that valuation in respect of the work within the work categories. The amount of the balance of the adjustable work included within that valuation is then multiplied by the fluctuation amount for the work categories divided by the total value of work executed in the work categories for that valuation. A simplified example using only two work categories is given in Table 20.2.

It will be recognized that the application of the formula requires the calculation on each monthly valuation of the total value of work executed within each work category. Since there are 38 of these which may be spread over a large number of bills of quantity this is no small exercise for the quantity surveyors. There is a slightly simplified method but it still does not get over the major hurdle of valuing the work each month by reference to

**Table 20.2  Example of the Osborne formula**

|  |  | £ |
|---|---|---|
| Contract sum |  | 100 000 |
|   Deduct provisional and PC sums including main contractor's profit | 20 000 | 20 000 |
| Total of contract price subject to adjustment |  | 80 000 |
|   Deduct work allocated to work categories |  | 70 000 |
| Balance of adjustable work |  | 10 000 |

Breakdown of work into work categories on interim valuation number 2:

| Category 1 | £5 000 |
|---|---|
| 2 | £15 000 |

CALCULATION OF ESCALATION

| Work category | Valuation | Month of valuation | Work category index Base | Increase |
|---|---|---|---|---|
| 1 | 5 000 | 120 | 100 | $5000 \times \dfrac{20}{100} = 1000$ |
| 2 | 15 000 | 140 | 120 | $15\,000 \times \dfrac{20}{120} = 2500$ |
|  |  |  |  | $3500$ |

Valuation of balance on adjustable work within the valuation is £2 000. Escalation on balance of adjustable work:

$$2000 \times \frac{3500}{20\,000} = \qquad\qquad 350$$

| Total escalation | 3 850 |
|---|---|

the selected work groups.

Specialist engineering sub-contracts for mechanical and electrical installation and lifts are dealt with by separate formulae.

There are similar provisions in the Osborne as in the Baxter formula for not paying increased escalation for the period during which the contractor is late in completing the works. It is also worth noting that under both formulae the amounts payable for escalation are subject to retention. The interpretation of both formulae is also that the value of unfixed materials delivered to site should not be included in the amount which is subject to fluctuation until they are incorporated into the works. It has been suggested that this could lead to an unjustified increase to the contractor if materials are brought on to site early, but against this there is an advantage to the employer in having materials brought to site in a manner which will assist in timely completion and the contractor is only paid a proportion of their value so that he is in part financing their cost.

On civil engineering and building works overseas the main problem in the assessment of price escalation is the lack in most territories of reliable and regularly published indices. Either they are unavailable or they are subject to political influence. Clearly separate formulae will need to be established for those portions of the works which are subject to UK or other overseas escalation and those which are subject to local factors. If for the latter there are no appropriate indices, it is suggested that the contractor has only the option of reverting to a list of basic locally procured materials and local wage rates and claiming the difference between these and those which he actually pays or possibly as regards wages to construct a site index as suggested earlier on site installation (see p. 360). Escalation is, however, in many territories politically unpopular on government contracts as it is seen as a reflection on the government's liability to control inflation, and even therefore if the contract does contain escalation provisions the contractor will not necessarily get the escalation paid when he legitimately claims. Although therefore there is an obvious risk in quoting a fixed price it may well prove to be the lesser of two evils.

# 21 Functions of architect/ engineer and purchaser

One of the distinguishing features of the forms of contract developed by the professional institutions in the UK for the carrying out of building or civil engineering works and the supply and installation of mechanical and electrical plant is the role given to the architect/engineer. From even a cursory look at the sets of contract conditions, it is apparent that in some ways the architect/engineer is, in a sense, an additional party to the contract along with the employer and the contractor. Why is there need for an appointment of this nature, what is its contractual significance, and how in practice does it work?

To answer these questions it is necessary to examine the duties which the architect/engineer is required by the terms of the contract to perform and to divide these into two groups: first, those which are basically administrative, where he is subject to the instructions of the employer; second, those functions which require the engineer to make decisions where he is required to form and act on his opinion, in which he is expected to act within the terms of the contract impartially, honestly and with professional integrity towards both parties. At the time when the first edition of this book was published the view which prevailed then was that in exercising the power to make such decisions the engineer was acting in the role of a quasi-arbitrator. That view can no longer be maintained following the decision of the House of Lords in *Sutcliffe v Thakrah and Others* 1974, but that does not alter the engineer's responsibility to act in an unbiased manner. This function may now be referred to as the engineer's 'independent role'. It is this second function which may be confusing when one is introduced to contracts of the above nature for the first time. The view expressed in the previous edition of this book, that the engineer could be held liable to the contractor for negligent certification, is now subject to considerable doubt following the decision in *Pacific Associates v Baxter* [1990] QB 993 (CA). The difficulty is that the structure of the contractual relationships and the existence of a wide arbitration clause provide the contractor with a remedy against the employer in contract for the default of his agent. Is it then reasonable to provide the contractor with a separate remedy against the engineer in tort? Generally it is considered, following the *Baxter* case, that the answer must be 'no', but not possibly in all cases. An engineer might know that any remedy in arbitration was in practical terms an illusion and that the contractor had entered into the contract relying wholly on the skill and probity of the engineer. (See the article by Duncan Miller in the *International Construction Law Journal* 1993, Lloyds of London Press, at p. 172.)

Before considering the powers and responsibilities of the architect/engineer in this connection in more detail, the following table dividing his duties under the contract into the two groups referred to above may be helpful.

## Duties of architect/engineer

### Under client's instruction
1 Furnishing the contractor with drawings and information.
2 Issue of variation orders altering extent, nature, or quantity of the works.

3   Suspension of the works.
4   Nomination of sub-contractors and suppliers.
5   Approval of the work and inspection.

### Independent role
6   Pricing of variation orders where new rates or prices must be established.
7   Pricing of additional sums which may be due to the contractor for suspension, unforeseen circumstances, etc.
8   Adjudicating on the validity of claims presented by the contract.
9   Granting of extensions of time.
10  Issue of certificates.

Let us consider first those duties which the architect/engineer performs acting under his client's instructions.

### Furnishing to the contractor of drawings and information

A main function of the architect/engineer is to act as the focal point for communication between the employer and the contractor. To ensure a single line of official communications between the parties is an absolute 'must', so on the one side we have the architect/engineer and on the other the contractor's contract manager or engineer. If duplicate lines of communication are allowed the only likely result will be misunderstanding, contradictions, conflicting instructions, and ultimately an administrative nightmare. This is not to say, of course, that the architect/engineer is permitted to act entirely on his own initiative in the exercise of this function. In so far as he is acting under his client's instructions, it is for the employer to establish such internal procedures and disciplines as he may consider necessary to ensure that the architect/engineer consults with the specialist functions in the employer's organization in other fields on those matters which are their concern. But with one exception to which reference will be made later, none of these people should be allowed to communicate directly with the contractor, nor is the contractor concerned whether the architect/engineer has in fact complied with the employer's internal procedures. All that the contractor has to ensure (and this can on occasions be difficult enough) is that, on those matters which are reserved by the contract to the architect/engineer, the contractor acts on the instructions of no one else, no matter how eminent they may be in the employer's organization, without getting such instructions confirmed by the architect/engineer in writing in the manner prescribed by the contract.

## Issue of variation orders

The same principles apply to the issue of variation orders. The employer will no doubt wish to limit the extent to which the architect/engineer is entitled to vary the contract without prior consultation. Such limitations may be expressed either by reference to the type of variation, or by imposing a financial limit both on the value of the individual variation order and the total sum which may be expended by the architect/engineer on variations. But again, none of this is of any concern to the contractor, who is entitled to act on the basis that any instructions issued by the architect/engineer under his powers, as expressed in the contract, are binding on the employer. Under clause 2(1)(b) of the ICE conditions, 6th edition, any restrictions on the engineer's authority are required to be set out in the appendix to the form of tender.

## Approval of work and inspection

The role of the architect/engineer in relation to the inspection of work and materials and to the approval of work as finished is a difficult one to define. He may in fact, under the same contract, be acting in these respects both under his client's instructions and also in his independent role. This may come about in the following way. When exercising the powers which are given to him by the normal clauses in the contract conditions on inspection during the course of manufacture or examination of work on site, the architect/engineer would be acting simply on behalf of the employer. Thus, although he should act reasonably as a professional man, his duty at that stage would be to the employer, to protect the employer's interests, and he would have no duty to the contractor to act impartially. He would be entitled to accept instructions from the employer as to the manner in which he was to exercise his powers.

In the event, however, of the contractor disputing the architect/engineer's decision, or at a later stage of submitting a claim that by reason of such decision he had been put to extra expense over and above that which he had reasonably contemplated when entering into the contract, or had been delayed in the execution of the contract, then the architect/engineer, now acting in an independent capacity, must decide on the merits of the contractor's claim, and in so doing must act fairly and impartially between the parties. For example, the engineer, as the employer's agent, may decide that the finish on certain concrete does not accord with the high standard which he knows that the employer wants, and may reject certain work and require other work to be proceeded with by methods of working which are slower and hence more costly than the contractor had estimated on. At the time the contractor may accept such a decision, but may subsequently put forward a claim for an extension of time and increased costs. In considering such claims the engineer must act upon a fair and proper interpretation of the contract as an independent observer.

So much for the duties which the architect/engineer performs under his client's instructions. We must now consider those functions which he performs in his independent role in which he acts according to his own judgement and opinion as a professional man.

## Pricing of variation orders and issue of certificates

It will be convenient to consider together those duties which involve the architect/engineer in certifying to the contractor the sums which he is entitled to be paid under the contract. The architect/engineer's duty here is clear: he must give the certificate on his own judgement and without any improper interference from the employer. In view of the extent to which even in the UK it has become increasingly common in recent years for employers to seek to influence or even direct architects/engineers as to the manner of the performance of their independent duties it is appropriate to re-state the position of both the employer and the architect/engineer.

In the absence of any express term in the contract, where a government servant is required to act as a certifier (in the case in question of extensions of time), then terms will be implied that government will not interfere with the duties their employee, as certifier, has and will ensure that he does in fact perform his duty as such (*Perini Corporation v Commonwealth of Australia* [1969] 12 BLR 82). Acts of the employer which would amount to obstruction or interference with the conduct of an architect when acting within the sphere of his independent duty would include directing him as to the amount for which he is to give his certificate (*Burden v Swansea Corporation* in the House of Lords [1957] 3 All ER).

The contractor is entitled to receive and indeed has to be able to rely upon that which he contracted to receive, the fair decision of the architect/engineer – who must not deliberately misapply the provisions of the contract with the intention of depriving the contractor of sums to which he is entitled (Court of Appeal in *Lubenham Fidelities v South Pembrokeshire DC*, 9 CON LR 85). At the same time it has been recognized judicially that notice must be taken of the interests of the architect/engineer as they will be presumed to have been known to both contractor and employer at the time of entering into the contract. These are:

1   The architect/engineer is an agent and in some cases a salaried servant of the employer and in consequence owes a duty to the employer for reward.

2   It is usual for the architect/engineer, before the contract is placed, to have made for the employer an estimate of the cost of the works. This gives him a certain interest in that estimate not being exceeded. Normally that interest will not extend to the point that, should the estimate be exceeded, then the fees of the architect/engineer will be affected. If it did, then it would seem arguable that the architect/engineer has been put in a position in which it is not possible for him to act in the independent manner which would normally be expected

of him, and that the contractor's attention should be drawn to the position at the time when tenders are invited.

3 The architect/engineer is under an obligation to his employer and has an inducement out of regard for his own reputation not to allow unnecessary extras, and to keep the cost of extras down to a reasonable level.

4 In the exercise of his duties as agent, the architect/engineer is in frequent communication with the employer and with the contractor. As agent for the employer he may be called upon to give the employer advice which as regards the contractor is of a confidential nature and not be disclosed to the contractor. When, however, he is acting in an independent role, the architect/engineer must endeavour not to communicate to one party that which he does not communicate to the other in relation to the subject matter of his duties. Thus, if an engineer prepares a report on the facts relating to a disputed item in an application for a certificate, such report should be made available both to the employer and to the contractor.

## Extensions of time for completion

When the contractor is delayed due to a cause which is beyond the contractor's reasonable control, then he is normally entitled to be granted such extension of the time for completion as the architect/engineer may consider reasonable. In deciding whether the cause of the delay was beyond the contractor's reasonable control and, if so, on the period of extension to be granted, the architect/engineer is again acting in an independent role. It would, for instance, be completely wrong if, knowing that completion on time was vital to the employer, he were to refuse to grant an extension of the time of completion to which he knew the contractor was entitled under the terms of the contract. It would be equally wrong if the employer were to give the architect/engineer instructions that under no circumstances were extensions of time to be granted. In both cases any decision of the architect/engineer on an application by the contractor for an extension would in law be a nullity, and the architect/engineer would be disqualified from so acting in this respect for the remainder of the contract.

It would of course, however, be recognized, as in the case of claims by the contractor for additional costs, that the architect/engineer, having given the employer initially an estimate of the time which it will take to complete the contract, has an interest in seeing that such estimate is not exceeded, and that he will therefore be expected to examine carefully any requests for an extension of time.

There will be many instances where the completion of one contract to time is vital to the successful completion of the entire project. If such a contract starts to run late and the architect/engineer advises the employer that the cause of this is one which under the contract entitles the contractor to an extension of the time for completion, the employer is entitled to reply in these terms:

Very well, I accept your decision as an independent observer that the contractor is entitled to a six weeks' extension of time. I am sorry, however, but my overall programme for the project is such that I cannot afford it. You now, as my agent, must negotiate with the contractor a revised bargain which will ensure that the lost time is recovered, and I am willing to spend up to £x for this purpose.

Further it is not open to the employer to give, or for the architect to accept, instructions from the employer which deprive him of his independence when certifying sums due to the contractor. So in one case an architect who had failed to issue a certificate when he should have done wrote to the contractor saying 'in the face of their (his client's) instructions to me I cannot issue a certificate whatever my private opinion may be'. It was held that the architect was disqualified and that the contractor was entitled to sue for the amount which should have been certified despite the non-existence of the certificate, so decided in *Hickman v Roberts* in the House of Lords [1913] AC 229.

So far in this chapter reference has been made to the position of the architect/engineer under an English form of contract administered within the UK. However the independent role of the architect/engineer is virtually unknown outside the UK despite the efforts of UK architects and consulting engineers. There is great danger therefore in the contractor assuming that he will be protected in the same manner overseas as he would be in the UK even when a UK consultant is employed. If a local firm of consultants or worse still a member of the employer's staff is nominated to exercise the powers of the architect/engineer, then the contractor can expect only that they will do so looking merely to the interests of the employer. Impartiality as between the parties is not a word they are likely to understand. Expressions such as 'in the opinion of the engineer' become a synonym for 'what does my employer want me to do'.

It must also be recognized that in the UK matters are changing. In describing the essential features of the NEC earlier (see p. 189) it was noted that there the traditional functions performed by the engineer are divided into four and it is essential that the person appointed as adjudicator is a separate person who is entirely independent of the parties to the contract. In arriving at this decision those who prepared the NEC recognized two problems. First, the extent to which employers today even in the UK wish to restrict the engineer's independence of judgement. Second, that in certain circumstances the engineer in his independent role will be called upon to adjudicate on disputes relating to claims for which he himself when acting under his client's instructions has been a party and may even be the cause and that this situation is no longer acceptable.

It is considered that this separation of powers between the management of the project and the adjudication of disputes will spread far wider than the NEC and will become the norm even in traditional forms of contract.

## Position of employer/purchaser

It is obvious from the description of the architect/engineer's duties given above that he is put in a position of great power and authority as regards both the contractor and the employer. What has to be recognized, however, is that the works are being built for the benefit of the employer, not that of the architect or engineer, and they are being built with the employer's money. Apart, therefore, from the specific powers which are normally reserved to him under the contract – for example, termination for default – the employer has a vital interest in the proper administration of the contract. For his own benefit, therefore, the employer should ensure that a proper system is laid down for the management and control of the contract, both technically and financially, and that is adhered to in practice.

The essential features of this system should be set out in the contract or terms of appointment of the architect/engineer on the following lines:

1   *Manager.* The person appointed to act as the employer's project manager should be named as the channel of communication between the employer and the architect/engineer.
2   *Designs.* The architect/engineer should be instructed as to any specific designs which the employer requires to approve in detail.
3   *Procedures.* Procedures for tendering for nominated sub-contracts or supply items should be agreed, with particular reference to any requirements of the employer's purchasing department as to, for example, standards, preferred suppliers, or bulk supply agreements.
4   *Restrictions.* Restrictions on the architect/engineer's power to issue variations without prior approval of the project manager. Restrictions should be defined in relation to:

   (a)   value of the individual variation
   (b)   aggregate value of all variations
   (c)   extension of time for completion
   (d)   effect on design of the works operating costs, and any other specific matters which are particularly significant to the employer.

5   *Programme.* The employer should be kept informed of the

**ERRATUM**

The first sentence of the fourth paragraph on page 377 should read 'Item (f) in the above list is most important.'

# Gower

*Contracting for Engineering and Construction Projects Fourth Edition* by P D V Marsh

physical progress of the work and should be consulted about any significant change in the programme and before any extension of time is granted to the contractor. When the architect/engineer proposes any variation to the works he should be required to declare whether it will have any effect on the programme or not.

6 *Cost reviews.* The architect/engineer should provide the employer with a regular report (monthly may be convenient) which shows:

(a)   original contract price
(b)   value of variations authorized
(c)   total current contract value
(d)   value of work completed to date
(e)   value of payments made to contractor
(f)   estimated value of contract to complete
(g)   under- or over-run on contract budget.

Item (f) in the above list is ~~not~~ most important. It can quite easily happen, for example, on a project with separate civil engineering and plant contracts, that a variation on one affects another, but the value of the consequential variation cannot be assessed until later date. Similarly, if the civil contractor is being paid on remeasurement and for any reason the quantities of work to be executed are likely to exceed the quantities shown in the bill, but the additional quantities are due not to a variation but to a change in ground conditions from those anticipated, this again may be known some time before the extra costs are incurred. The employer must have early warning of events of this sort, and he gets it through the estimate of what it is going to cost to complete. Also, if the architect/engineer has knowledge of a pending claim, which he knows is in part justified, it should be shown here.

7 *Certificates.* The architect/engineer should provide the employer with copies of the certificates as they are issued so that he can approve payment of the contractor's invoice when received. The architect/engineer should consult with

the project manager before the issue of the completion or taking over certificate and again before the issue of the final certificate, so that, for example, the views of those departments in the employer's organization which will be concerned with using the works can be made known to the architect/engineer before the employer is committed by the certificate being issued to the contractor.

8   *Claims.* The architect/engineer should advise the employer of any claims submitted by the contractor and of his proposals for the settlement of these.

# 22 Variations in price and time

Variations may not unfairly be described as the cancer of contracting. In quantity their cumulative effect can operate to destroy the best of contracts: the habit of ordering them is in itself a disease. What causes this disease? The causes are many but the principal ones may be summarized as follows:

1  Inadequate allowance for thinking time. It is distressing but true that many managements are still not convinced that progress is being made unless holes are being dug on site or plant manufactured.

2  Inadequate specifications. One finds a great reluctance amongst people to be completely specific as to what they require, as to the services which the employer will himself provide or the actual conditions under which the work will be carried out.

3  Insufficient attention paid as to whether what the tenderer is offering is in fact precisely what the purchaser wants to buy. The tendency to say 'That's a matter of detail we can sort out later'.

4  Lack of discipline. In the matter of variations it is often far

easier to say 'Yes, while we are about it we might as well have that done' than to say firmly 'No, it's not necessary'.

5   Improvements to avoid obsolescence. With the rapid rate of technical change taking place today any major plant is likely to be out of date in some respects long before it is completed. There is always the temptation to try to avoid this by incorporating improvements in the design.

6   Genuinely unforeseeable circumstances. It would be idle to pretend that no variation is ever justified. There will be times when conditions do arise when it is essential to vary the works – for instance, the existence of unsuspected drains or cables which have to be diverted.

What is often not fully appreciated is the effect which even quite a simple change of specification can have on a contractor. This may involve him in:

1   Design work which because of the change is now abortive.

2   Additional design work including studying the consequential effect of the variation on a number of drawings.

3   Cancellation of, or modification to, orders already placed on his own works or on outside suppliers.

4   The placing of new orders.

5   Delay and/or rephasing of the manufacturing programme to accommodate the variation.

6   Delay in delivery of material to site due to action under 3 above.

7   Rephasing of site works or concentration of work into a shorter period with consequent additional overtime costs and loss of productivity.

8   Extending the period to the contract.

It follows from the above list that unless the variation is ordered very early in the contract indeed, the assessment of the effect of the variation either in terms of cost or time is not easy. Consider first the question of the assessment of the change in the contract price for a plant due, say, to the deletion from the specification of one item and the substitution of another.

Table 22.1 represents the direct financial balance between the item originally included and that now ordered as a variation. It takes no account of the factor of time. Taken in isolation this is correct, unless the single variation itself is so great that it does have an immediate effect on the overall programme. It also takes no account of the double administrative cost effect on the contractor of having to go through the same operation twice. The contractor's staff, whose services are recovered for under the estimate as a percentage of prime cost, will have been involved to some extent on the item already in estimating and procurement, but under this listing the contractor would recover for such services only once for the new item. Again, if it is only one item, few contractors would seriously quarrel with this, accepting it as one of the hazards of contracting. The trouble starts when it is not one variation but a series of variations, when

**Table 22.1  Financial balance between item originally included and that ordered as a variation**

| ADDITIONS | DEDUCTIONS |
| --- | --- |
| Works or bought-out cost of the new item. | Works or bought-out cost of the item to be replaced. |
| Percentage for overheads and profit related to works or bought-out costs. | Percentage for overheads and profit related to works or bought-out costs. |
| Man-hour costs for installation of new item. | Man-hour costs for installation of the item to be replaced. |
| Percentage overheads and profit related to installation costs. | Percentage overheads and profit related to the installation costs. |
| Charges for additional design work including overheads and profit necessary to incorporate new item. | Charges for any detailed design work which will no longer be required including related overheads and profit. |
| Design, labour, and material costs and related overheads and profit on any consequential modifications or alterations to the remainder of the plant, including study of drawings to determine whether any such are necessary. | |
| Cancellation charges payable to outside supplier or costs or any work actually carried out in contractor's works. | |

the programme is affected, and when the time spent by the contractor's head office starts to become totally disproportionate to the value of the contract. Under these circumstances the employer must expect that the contractor will seek to recover additionally for:

- abortive time spent by head office staff not otherwise directly charged to the contract
- prolongation of the contract period on site – for example, hire of huts, supervisors' salaries
- loss of productivity and overtime working due to changes in the programme.

It is easy enough to set down the basis on which single variations should be priced in the manner which has been done above. It is often, however, another matter actually to negotiate the alteration in price. The purchaser will be thinking the contractor is trying to take him for a ride, but may additionally be genuinely unappreciative of what trouble and cost his simple instruction has caused. He will also be acutely aware that he cannot get competitive quotations. The contractor may be anxious to recover some of the ground he lost in post-tender negotiations. Neither side is likely to be in the mood for concessions, but the purchaser will probably be in the weaker bargaining position.

Partially for this reason attempts are sometimes made to establish in advance the main tender rates on which variations can be calculated. It is possible to do this for civil engineering or building work or for structural steel or pipework, although the value of doing so seems questionable. This is because in putting forward his rates the contractor must make certain assumptions regarding the quantity and complexity of work which will be involved, the plant required, and so on, and as to whether it will be convenient to do the work in parallel with or as an extension of existing work of the same nature; or whether it will be something quite separate for which perhaps plant and a gang of men must be specially brought to site. For this reason, and also because it is difficult to take rates for the purpose only of pricing variations into account in deciding on the award of the contract, the tenderers have every incentive to assume the worst conditions and price accordingly.

In general therefore it would seem preferable from the purchaser's point of view, despite the difficulties involved, to negotiate when the occasion arises and on the facts of the particular variation without being tied in advance. The contractor may, however, press, for quite a different reason, for at least the overhead percentages and margins to be fixed and stated in the contract.

It is often assumed that contractors welcome variations in that they can use them to recoup any losses they may have made on the main contract or at least improve their overall rate of recovery on the job. While, as explained above, the contractor may be placed in a favourable negotiating position when it comes to settling a price for the variation, it has also been pointed out that the cumulative effect of a number of variations on his main contract programme can be extremely serious and result in disruptions of work, loss of productivity and so on. These losses, while real, may often be difficult for him to quantify or to claim from the employer. In any event he is likely to be involved in protracted claims negotiations which are both time- and cost-consuming in themselves and may well be detrimental to his chances of obtaining further business from the employer concerned.

For this reason some contractors seek to put forward as part of their tender, rates or percentage charges for different classes of work which may be involved in handling variations – for example, design, which are deliberately so high as to be penal. In this way the contractor seeks to utilize the contract as a means of disciplining the employer's engineers.

While obviously such an arrangement can be open to abuse, there does seem considerable merit in any system of pricing which will bring home to those responsible for administering contracts the real cost involved in having frequent changes of mind. Accordingly a system of differential pricing for work as a variation as compared with the same work under the main contract seems justified. If as a result variations become a luxury which can be afforded but rarely, then so much the better. It might also help to avoid the other practice, of including an allowance within the original tender for the 'messing about' which, from past and often bitter experience, the contractor knows that he is likely with certain clients to receive.

The NEC adopts a different approach in that it requires firms as part of their tenders to submit a schedule of cost components – labour rates, plant rates, design charges, overhead percentages together with a percentage fee. The employer includes in his enquiry provisional amounts for each of these and the sum total is taken into account in the tender comparison. These rates and percentages are then used in the assessment of compensation events which includes variations.

Of course these rates and percentages are only half the story since there still remains the issue of the quantities to which they are to be applied and the productivity factors involved.

However, with the NEC it is important to note the principle that compensation events are priced on the basis of the actual or estimated change in cost incurred by the contractor, in the latter event using the schedule of cost components and fee percentage, and not by using the rates and prices for work in the contract from which the original contract price was derived.

A vital factor in the successful control of variations is the timing of price negotiations. Only too often because of the pressure for physical progress with the work and the complexities in assessing the price change, instructions are given to the contractor to make the change, with the alteration in price to be negotiated later.

Ideally the sequence of events should be:

1 Purchaser decides that a particular variation would be desirable.
2 Contractor is instructed to assess the effect of the proposed variation in terms of :

- price
- time
- performance.

3 Contractor submits his proposals under the above three headings.
4 Purchaser decides whether he can afford the variation taking all factors into account.
5 If purchaser decides to proceed with the variation, then he negotiates amendments to price, time for completion and specification.
6 Purchaser issues formal variation order in writing, using a standard serially numbered form.
7 Contractor proceeds with the work.

This seems a long series of steps; the temptation is there to go straight ahead and tell the contractor to start work. Indeed there will be genuine emergencies when it is necessary to do just that and tidy up the paperwork afterwards. But in doing so not only is any possible negotiating advantage lost, but also any curb on the enthusiasm of the purchaser's staff to make variations is removed and financial control of the contract is made impossible. Except in the case of a real emergency it should be made difficult to make variations.

The procedure referred to above is essentially that which has been adopted in the NEC.

However, while it may be possible at the time to assess the direct effect of the individual variation on the contract price and time for completion, it is much more difficult to assess the indirect or consequential effect. This with one variation may be small, but as the number of variation orders grows so do the consequential effects increase, often at a much faster rate.

While therefore, ideally, one should treat each variation order separately and assess finally its effect on the contract price and time before it is issued, there are occasions when it is just not practicable to do this. In order to retain as much control as possible in these circumstances it may be necessary to divide the negotiation of variations into two stages:

1 The assessment of the direct effect of the variation.
2 The assessment of the consequential effect of the variation on the contract price and the overall time of completion.

Stage 1 should be completed for each variation order before it is issued. Stage 2 cannot be completed until the design has been finally frozen. At that point the cumulative effect of the variation orders can be reassessed and any necessary adjustments to the contract price and programme made. Obviously the earlier the design-freeze date, and so the final contract value and programme, can be established the better for both parties. What is vital, however, to do at the time is to record and agree with the contractor the facts on which the stage 2 negotiations will be based. There is no excuse for there not being accurate records of, for example, the time plant was on site and the periods during which it could not be fully utilized.

Not all variations relate to the physical content of the works. The employer may wish either to speed up completion or to slow it down, or possibly to put the contract into suspense. Any such actions are bound to have a serious effect on the contract price.

The simplest case is probably trying to speed up completion. Time may be bought by:

- working additional overtime or at weekends
- putting on an additional shift
- offering suppliers or sub-contractors a bonus to deliver or finish earlier.

By such methods small improvements can be obtained fairly easily. But above quite a low level the law of diminishing returns starts to operate and it becomes more and more expensive to purchase smaller and smaller improvements. Once a certain level has been passed the productivity value starts to drop rapidly, and on double shifting the productive effort may be 25 per cent or more below normal. Moreover, the longer one tries to continue with excessive overtime or double shifting, the lower the return one obtains for the increased expenditure.

As regards pricing, provided the make-up of the labour charges already included within the contract is known, this presents no real difficulty. For site work the make-up will normally comprise:

1   Basic wage which may in these days bear no relation at all to the so-called basic wage agreed nationally between the union and the employer's federation concerned.
2   Bonus often related to productivity.
3   Condition money which may cover such things as working in dirty conditions, wearing rubber boots, etc.
4   Subsistence allowance for men lodging away from home or radius allowance for those living within a certain distance from the site.
5   Travelling time.
6   Allowance for overtime. It is virtually impossible today to obtain site labour without a guarantee of a certain number of hours overtime a week.
7   National insurance, holidays with pay and common law insurance, all of which bear a direct relationship to wages costs.

To these the contractor will add his charges for supervision, small tools and consumables and other erection on-costs including normally a margin to cover his head-office erection department.

One important point to ensure, when negotiating an addition to cover for extra overtime, is that where such an addition is to be charged on a percentage basis, such percentage is charged only on those costs which are directly proportional to wages, or alternatively that the percentage is adjusted to take account of non-variable items. Item 4 in the above list, for example, is a flat weekly charge which will not alter.

Slowing down a job is rather more difficult, in that it will involve the contractor being engaged for a longer time on the contract and will therefore tie up his resources for a longer period, so reducing his potential earning capacity over that period. For this reason the contractor may reasonably claim under the following headings:

1   Charges for plant, huts, etc., retained on site for an extended

time.

2   Salaries and overheads of supervisory staff so retained.
3   Some additional charge for wages costs due to less productive work.
4   Additional costs for any work which is now to be carried out under different and more arduous conditions, for example excavation to be carried out in the winter instead of the summer.
5   If the contract is on a fixed price basis an addition to cover:

- any increase likely to be met in the extended period
- the proportionately more serious effect which increases occurring earlier in the contract period will have, over the allowance made for these when the estimate was prepared. For example, 40 per cent of the contract work may now be carried out after the date when a wages award will take effect, instead of the 25 per cent on which the estimate was based.

6   Additional interest charges due to retention moneys being outstanding for a longer period.

Where the contract is put into suspense, consideration will need to be given by the buyer to the following points:

1   Should the contractor's site organization plant, huts, etc., be removed from the site? Obviously, if all or any part of it remains, the contractor is going to want to be paid for it. On the other hand the costs of taking it away and then re-establishing it may also be heavy. The buyer must weigh up the advantages of each course, taking into account the likely period for the suspension.
2   Work partially completed on site must be properly protected; loose items not yet incorporated or built into the works must be identified, labelled or marked, and properly stored. If the contractor's organization is being removed from the site then the responsibility for such storage and safe custody will vest in the purchaser.

3   Items in course of manufacture or not yet despatched must be similarly treated. In this case, however, they should remain at the risk of the contractor; this needs making clear explicitly; also the buyer will want to make sure that the contractor has insured the items against all insurable risks.

4   The contractor will seek to ensure that he is not prejudiced by the suspension as regards the time when payments under the contract should be made. Thus if the contract provides for retention money to be released on completion, and completion is delayed as a result of the works being suspended, he will want to be paid the retention moneys relating to work already executed not later than the date by which they would originally have been released. This is reasonable, and certain standard conditions of contract do make provision for this. It is also reasonable to make payments on account of work partially completed in the contractor's shops but not yet delivered or ready for delivery, provided that it has been identified as the purchaser's property. The buyer will want to make sure that such parts are correctly marked and so on, and that they are covered by all-risks insurance.

5   From the buyer's point of view it also seems reasonable that he should not as a result of the suspension lose the rights he may have in respect of any defects which may occur in the works after they have been finally completed. In other words, payment of retention moneys in respect of the partially completed job must be without prejudice to the defects liability period, which should only start to run after the actual completion of the job. Where, of course, equipment which suffers natural deterioration no matter what care is taken is stored for any period, this must be subject to the contractor's right to inspect and make good the results of any such deterioration.

## Variations on overseas contracts

Variations, unless they are of a minor nature which can be accommodated with existing resources and the contract programme, are even more troublesome when the contract is being executed several thousand miles from the contractor's home base.

First, there is the problem that the design or planning for the variation may have to be referred back to the UK or personnel sent out from the UK specially to site for that purpose which all takes time and costs money.

Second, the contractor's site organization and facilities will have been geared to the contract as it is known and will lack the flexibility for adjustment which is possible within the UK. A large variation will therefore have that much greater impact on the economic utilization of existing resources – plant and labour may have to be retained even if there is no immediate use for them – and additional resources of a different character which may be required will take time to have available.

Third, as has been indicated earlier the cost of retaining supervisory staff on site is that much higher because of the expenses of housing, feeding etc. so that the labour costs associated with the variation will be substantially increased.

# 23 Claims and their negotiation

It may well be asked by someone coming new to contracts for construction works why it is that the subject of claims, and what is often referred to as 'claimsmanship' by contractors, should occupy such a prominent place both in the literature on standard forms and their practical administration. The main reasons are:

1   The very nature of constructional contracts carried out, as they largely are, on open sites and with the uncertainties necessarily attached to works involving excavation below ground.
2   The division of responsibilities between the engineer/ architect and the contractor under the traditional methods of contracting as described in Chapter 2.
3   Failure of pre-contract planning both by the employer with the advice of his engineer/architect and by the contract in the preparation of his tender. This is due largely to an unwillingness to spend the time and money necessary for proper investigation of site conditions and construction methods, to provide the firms tendering with the fullest information on the engineer's/architect's intentions

regarding design and allow an adequate time for tendering.

4   Failure on the part of the engineer/architect to obtain adequate information at the time of tendering as to the contractor's proposed methods of construction and programme for the carrying out of the works and to compare this with his own intentions so as to satisfy himself about their compatibility.

5   Inadequate attention paid to the pre-qualification and selection of firms to be invited to tender and to the analysis of their bids, not just in relation to the overall price but to all other data required to be submitted.

6   Extensive variations ordered during the contract period. It is interesting to note that many of the cases arising on this issue have concerned buildings such as hospitals, the design of which has clearly been subject to substantial post-contract alteration as a result of changes in operational requirements. This is due, one suspects, to a failure at the planning stage properly to involve those who would ultimately have the task of using the building for their professional purposes.

# Types of claim

Claims can be divided into four categories:

- claims for the payment of damages due to the employer's breach of contract
- claims for additional payments under specific provisions of the contract
- claims arising out of variations
- claims for disruption and delay.

### Claims for the payment of damages

The basic principle upon which any such claim must be founded is the same as that which applies to any other claim for damages, namely that the claimant is entitled, once the breach has been proved 'to be placed, so far as money can do it, in the same position as he would have been had the contract been performed'.

It follows from this that if the contractor can establish that, as a result of some failure by the employer to comply with his obligations, the contractor has suffered additional costs then he is entitled to recover these. Further, if the employer's failure is such that the contractor has been compelled to carry out work additional to that which he had undertaken to do under the contract, in order to enable him to comply with his contract, then he would be entitled to claim for additional profit on such extra work. It is not, however, the case where the contract is continuing that a claim for loss of profit can be made merely because some additional expense has been incurred as a result, say, of the contractor being delayed in the performance of the work as a result of the employer's default. For such a claim to succeed it could only be on the basis that the delay had been so prolonged, and the contract so substantial a part of the contractor's business, that it had tied up his resources to the point at which he had lost the opportunity of tendering for other potentially profitable business. This point will be considered further in the section claims for disruption and delay.

The situation is, however, different where the contractor's claim arises on the contract being terminated. In the case of *John Jarvis v Rockdale Housing Association* 10 CON LR the contractor terminated under clause 28.1.3.4 of JCT 80 and clause 28.2.2.6 provides that upon such termination the contractor shall be paid inter alia 'any direct loss or damage caused to the contractor as a result of the termination'. In the course of giving the judgement of the Court of Appeal Lord Justice Bingham said: 'The learned judge was content to assume that this clause gave the contractor the right to be paid all the profit that he would have made if he had completed the works in accordance with the contract and before us neither party challenged that assumption'.

The rights of the contractor to claim damages, and in particular to claim for loss of profit, may be affected by the express wording of the contract. In this respect the 6th edition of the ICE conditions has made a number of changes from the 5th edition. The term cost is still defined in the same manner as before to exclude profit. However, in a number of clauses it is specifically stated that there is to be added to the additional costs 'a reasonable percentage addition in respect of profit'. See, for example, clause 12(6) dealing with additional costs due to adverse physical conditions and artificial obstructions; clause 42(3), delay by the employer in giving possession of the site; but only additional cost is to be paid under clause 7(4), delay by the engineer in the issue of drawings.

In this respect one can contrast JCT 80 clause 26 which expressly provides that the provisions of clause 26, dealing with the contractor's right to the recovery of loss and expense caused by matters materially affecting the regular progress of the works, is 'without prejudice to any other rights and remedies which the contractor may possess' and so leaves open the contractor's rights to a claim in damages for breach of contract. In practice, it may not often be necessary for the contractor to invoke such a right. The expression 'direct loss and expense' has been interpreted by the courts as meaning the loss or expense which arises naturally and in the ordinary course of events, i.e. the damages recoverable under the first limb of *Hadley v Baxendale*.

### Claims for payment for compensation under express terms

As already indicated, most standard forms of contract do provide that in particular circumstances the contractor is entitled to submit a claim for compensation. The most obvious ones are where there has been a delay by the employer, or more likely the engineer/architect acting on his behalf, in carrying out their respective obligations. The basis upon which such claims should be made is by way of a comparison between the costs which the contractor reasonably expected to incur and the increases which he did in fact incur *arising out of the delay*. The practical difficulty with making any such comparison is that only too often the evidence available is not convincing. Even if the contractor uses a sophisticated computer-based programming system, it is unlikely that any such system will, unless specially set up for the purpose, distinguish between those delays which are due to the default of the engineer and delays which arise due to other causes. However, there is no doubt that the better the programming methods employed and actually applied in practice, with regular updating and identification *at the time* of 'holds' which have occurred and corrective action being taken, the better the chances are of a reasonably negotiated solution without the expense of protracted legal proceedings. Unfortunately only too often both sides see it as being in their best interest 'to play their cards close to their chests'. Even when the contract provides for the submission of regular programmes to the employer those supplied are more likely than not 'political' programmes produced for the purpose of either keeping the employer happy or providing the grounds for a subsequent claim, rather than being the true programmes to which the contractor is working. Contractors may not keep two sets of books for the purpose of defrauding the Inland Revenue but they most certainly on many projects keep two sets of programmes. Indeed it is not unknown for there to be three; one for the client, one for head office and one for site!

If the contractor is going to rely on being able to base a claim on an express term of the contract then it is essential for him to have complied with any procedure which is established in the contract for the giving of notices within the time-scales prescribed. This is an area in which contractors are notoriously lax. Partially, this is simply poor administration by people more concerned with the immediate problems involved of getting things built, and partially it is due to a not entirely unjustified fear of upsetting those on the employer's side particularly at site level upon whose co-operation the contractor is dependent for achieving results. It must always be remembered that any claim which is based on an allegation of employer default can and often will be looked upon by the recipient as a personal criticism. But no relationships however good which have been established locally during the course of the contract will prevent the lawyers or other professional advisers acting for the employer in dealing with the claim from taking the point, if such is the case, that notices were not given in due time and this could well be fatal to the success of the claim, whatever its other merits.

## Claims arising out of variations

The question of claims arising out of a multitude of variations which create the problems of prolongation and delay will be dealt with in the next section. Here it is proposed to consider two points:

First, the pricing under a re-measurement type contract of changes in quantities which are not covered specifically by variation orders. The point arises in the following way. According to the conditions of contract in general use in civil engineering (the ICE Conditions 6th edition and FIDIC 4th edition) the amounts to be paid to the contractor are to be determined in accordance with the admeasurement of the quantities of work actually executed and the quantities stated in the bills on which tenders were invited are only estimates. It can, and quite often does, happen that the actual quantities in respect especially of items involving excavation exceed by a substantial margin the quantities stated in the bills. In those circumstances the contractor will consider, not unreasonably, that the bill rate should no longer apply since the time to carry out the work and even the methods of construction involved may vary substantially from those which he contemplated when he planned his tender. Accordingly, as referred to earlier, the ICE conditions now provide specifically that, if the engineer so considers that the change in quantities warrants it, he shall after consultation with the contractor establish a new rate. At one time it was considered - see I.M. Duncan Wallace, *Construction Contracts*, 1986, Sweet & Maxwell, p. 113 - that the matter had been conclusively decided, as regards the FIDIC conditions, the other way round, as a result of the decision by the Court of Appeal in South Africa in *Grinakar v Transvaal Authority*. Fortunately in my view this position has now been reversed as a result of the Privy Council decision in *Mitsui Construction Co. Ltd v Attorney General of Hong Kong*, 10 CON LR 1, where it was decided that the engineer did have jurisdiction to fix a new rate for any billed item where he was of the opinion that the differences between the billed and measured quantities of work made the billed rates unreasonable or inapplicable, regardless of whether there had been a formal variation order, an engineer's instruction in relation to the specification of work to be executed or simply a substantial difference between the billed and measured quantities.

Second, the extent to which a contractor can claim in respect of a variation ordered by the engineer/architect that it is outside the scope of his authority. That position could arise in the following circumstances:

- if the engineer/architect were to order that work intended by the contract to be performed by the contractor were to be awarded to another party. Even the words 'that the architect has the power to give written directions as to the omission of any work' have been held in Australia not to entitle the architect to take away from the contractor and award it to a third party (*Carr v J.A. Berriman Ppty Ltd*, 1953 ALJR 273). It is arguable that the power to omit work applies only if the work is not to be done at all - see Max Abrahamson, *Engineering Law* and the ICE Contract 4th edition at p. 172 and the Irish authorities there stated.

- if the variation ordered was of a kind which significantly changed the nature of the works or required the contractor to undertake work of a significantly different type from that contemplated by the contract. The argument here is based on the premise that the power to order variations is based on these being necessary or desirable in relation to the contract works. The variation clause is not as it were 'a blank cheque' under which the employer can elect to have carried out under the terms of the contract other work which he would like to have done but which has no real relationship to that contemplated when the original contract was placed.

- when the payment basis of the contract is such that to require the contractor to perform variations without limit would place upon the contractor an intolerable burden and place him in effect entirely at the employer's mercy - *Sir Lindsay Parkinson & Co. Ltd v Commissioner of Works*, 1950 All ER 208.

### Claims for disruption and delay

One of the most common claims by contractors is that the number of variations ordered by the architect/engineer and/or the delays in the issue by the architect/engineer of the drawings and other information necessary to enable the contractor to proceed, are such that it is impossible to determine the effect of any one particular loss and that therefore the claim should be dealt with on the basis of the contractor's total loss on the contract. The obvious advantage to the contractor if he can bring himself within this ambit is that he does not have to prove details of each individual loss. He can apply a 'broadbrush' to the calculations and is unlikely, especially at arbitration, to come out with less than around 25 per cent of his original claim.

Equally the obvious disadvantage to the employer is that he does not have precise particulars of the sums being claimed nor of the basis upon which, in each instance, it is alleged that he, rather than the contractor himself or some external cause, is responsible for the loss in question. His ability to challenge the contractor's allegations is accordingly that much reduced.

So far as English law and practice is concerned there is authority for a 'total loss' claim being allowed where it is wholly impractical because of the complex nature of the interacting elements of the claim to consider these in isolation one from another: *J. Crosby & Sons Ltd v Portland UDC* [1967] 5 BLR 121. Since that decision there have been other cases in which the validity of a total loss or global claims has been doubted. In *Wharf Properties Ltd and Another v Eric Cummins and Associates* 1991, the plaintiffs argued that it was impossible to isolate specific areas of delay due to the complexity of the project. However, the Privy Council held that the claim put forward in which no attempt had been made to link cause with effect could not be allowed to stand. It was said 'The failure even to attempt to specify any discernible nexus between the wrong alleged and the consequent delay provides "no agenda" for the trial'. In *ICI v Bovis Construction Ltd and Others* 1992, again there was a failure to link the alleged financial consequences with each breach. The court did not reject the claim but required that ICI should prepare their claim in more detail, giving particulars of which clause of the contract had been breached and the alleged factual consequences of that breach.

It appears therefore that the position today is that the contractor must be able to show that he has made every effort practicable to itemize causes of delay and their individual effects. Only where because of the complexity of the inter-relationship between a number of causes is such that this is impractical is it likely that a court or arbitrator will accept a global claim.

In this connection it is now possible by the use of modern software to use the technique of 'impact analysis'. This establishes the impact of individual causes on a series of logically linked events within the network. However, the practical application of the technique requires the knowledge of how the work was programmed, how it progressed, when the delaying events occurred and the interaction between one delay and another. This again emphasizes the need for genuine contemporary data.

One of the many difficulties which are to be found in the calculation of claims for prolongation and disruption is that of head office overheads and loss of profit. There are in fact two possible bases of claim and it would appear that they are easily confused. First, there is a claim for overheads only which is based on the additional managerial time and expense which is required to deal with the problems which created the claim in the first place and seek their solution. It was to this which Mr Justice Forbes was referring in *Tate & Lyle v GLC* [1982] 1 WLR 149 when he said that to establish such a claim there must be evidence of the actual additional managerial time expended and he was not content to apply an arbitrary percentage. Second, there is the claim for both head office overheads and profit which is related to the fact that by virtue of the contract period being extended the contractor will be deprived over that period of the opportunity of earning a contribution to his fixed costs and of profits. It has become customary in the building industry to calculate such loss by means of a formula in order to avoid the need for detailed calculation and the one most commonly used is the so-called Hudson formula on p. 599 of the 10th edition of

*Hudsons Building Contracts.* While perhaps a little crude, it is simple to apply. An alternative is the Emden formula published in *Emdens Building Contracts & Practice.*

# Claims presentation and management

There are a few basic rules to be followed in preparing for and presenting claims. These are:

1 Consider the possible areas for claims from the start of the contract and plan accordingly. Don't wait until they happen.
2 Keep accurate records from the start of the contract – in particular a good, factual site diary.
3 Where it is considered that a claim may arise in respect of design work, ensure that the records are such that it is possible to trace the number of man-hours spent on revisions to each drawing and the particular reasons why such revisions became necessary.
4 Make a record of the requirements for the giving of notices and ensure all staff concerned are made aware of these.
5 Ensure that all correspondence with and from the employer which could have an impact on claims is reviewed, as are all minutes of meetings. Aim to answer allegations factually and as far as possible always 'put the ball into his court'.
6 In presenting the claim, make sure that it contains:

   ● a short executive summary
   ● clear references to the terms of contract on which the claim is based
   ● all essential data required in order to understand the claim, e.g. critical dates, extensions of time applied for and granted, variation orders issued, etc.
   ● copies of the programme, minutes and other documents supportive of the claim.

Perhaps the most difficult problem which the contractor faces in the negotiation of claims is the time which it takes. The employer has the contract works, the money and little inclination or incentive to part with them. Until recently employers were supported in this attitude by the ancient and much criticized rule of English Law established by the House of Lords that financial damages in the form of interest for late payment were not allowable. That position has been partially rectified by statute in that the courts may now award simple interest under 5.35 of the Supreme Court Act 1981 where payment is not made before proceedings are commenced. More importantly as regards contracts which contain such words as 'direct loss or expense' to describe the sums payable to a contractor in respect of the employer's default - see JCT 80 clause 26 - the Court of Appeal has held that such words cover the interest charges which the contractor has had to pay by being out-of-pocket. Such interest charges will run until the date of the last application before the issue of the certificate which relates to the primary loss concerned. Further, since such charges are truly in a contractual sense a loss suffered and not 'interest' they will be calculated on the basis charged by the contractors' bank, i.e. on compound interest with rests.

However, if a contractor is to succeed in claiming interest as part of the loss/expense incurred there must be some reference to that effect in the notice which he is required to give the architect under the terms of the contract: *FG Minter v Welsh Health Authority Technical Service Organisation* [1981] 13 BLR1 and *Rees and Kirby v Swansea City Council* 5 CON LR 34.

# APPENDICES

# Appendix 1 Draft instructions to tenderers for a plant contract

1 (a) You are invited to tender for the [*insert description of work*] at .............. in accordance with the attached Form of Tender.

 (b) The closing time for the receipt of tender will be 12 noon on ..............

 (c) You are required to submit .............. copies of your tender.

 (d) The tenderer is to acknowledge receipt of this invitation to tender to the purchaser's representative by fax immediately upon receipt and similarly to confirm within 7 days of receipt that he will be submitting a tender.

 (e) All requests for clarification must be submitted by fax or letter to the purchaser's representative and received by the purchaser no later than 15 days before the tender return date. Responses to clarification may be circulated in the form of an inquiry addendum.

 (f) The purchaser's representative for this inquiry is .............. to whom all correspondence should be addressed.

2  (a)   You are invited to tender on the basis that you will be responsible for the [*insert summary of contractors' responsibilities*] of the whole works defined as such in specification number ............... dated ...............

   (b)   A general description of the works is given in the attached specification which includes a statement of the duty which the plant is required to perform.

   (c)   If you should wish to submit for consideration an alternative or other variation, you must first obtain the purchaser's permission in accordance with paragraph 11. A statement of the salient features must be submitted with the tender for the alternative or other variation proposed. You shall, notwithstanding the submission of an alternative or other variation, submit a tender based on the specification attached.

   (d)   If a tenderer wishes to submit a tender in joint venture with another firm he must apply to the purchaser for permission to do so not later than ............... days prior to the date for the return of tenders and provide to the purchaser such full details of that other firm as the purchaser may require. Any consent given by the purchaser will be conditional upon the firms comprising the joint venture undertaking joint and several liability to the purchaser for the performance of the contract and to the inclusion by the firms with their tender of a signed copy of their joint venture agreement. [*This assumes that the firms have not been pre-qualified as a joint venture.*]

3  Your tender is to be submitted in accordance with the conditions of contract entitled ............... dated ............... [*copy attached*]. If you wish the purchaser to consider any modification to these conditions you must give full details of this in your tender. No undertaking is given by the purchaser that any modification requested by you will be accepted.

4  (a)   Your tender should be accompanied by a detailed specification and drawings sufficient to describe fully your offer. This should be set out so as to fit in with the sections into which your price is to be broken down as

given in Part 2 of the Form of Tender. You are required to complete the relevant section of the schedules to specification number ...............

(b) Your attention is drawn to schedule ............... to the specification in which you are required to enter the minimum numbers and categories of personnel which you consider would be required to operate and maintain the works efficiently.

5 You must provide with your tender [*here list any documents, drawings or other data which the tenderer is required to provide*].

6 (a) You are required to submit your tender on the basis of [*insert here whether tender is to be with or without price escalation; if with price escalation, the basis on which this is to be allowed should be stated in Part 4 of the Form of Tender*].

(b) Your tender must remain valid for a period of ............... months from the date on which it is due to be returned to the purchaser.

7 Evaluation of the tenders will be carried out by the purchaser using the following criteria:
[*Here list the criteria preferably in descending order or priority. This is an essential requirement if the contract is subject to the EC Procurement or the Utilities Directives and the selection is to be made on the basis of the most economically advantageous offer. However it is a good practice to adopt in all cases.*]

8 Whether your tender is accepted or not, you shall treat details of the specification and the documents attached hereto as private and confidential and in the event of a tender not being submitted the specification and drawings shall be returned. Any drawings issued to you are intended to be typical of the works to be executed and shall not be used as working drawings.

9 No tender shall be deemed to have been accepted unless such acceptance shall have been notified to the tenderer in writing by or on behalf of the [*insert official authorized to accept the tender*].

10 The purchaser does not bind himself to accept the lowest or any tender. On acceptance of a tender by the purchaser, the

successful tenderer may be required to enter into a formal agreement for the proper fulfilment of the contract.

11 The purchaser will not be responsible or pay for any expenses or losses which may be incurred by you in the preparation of your tender.

12 The tender and accompanying documents filled in as directed must be sent under cover of the 'tender' label accompanying this invitation to [*insert name of official concerned*].

13 Requests for permission to visit the site should be made to [*insert name of local official concerned*].

14 No alterations should be made to the Form of Tender all the blanks on which must be filled in.

15 (a)  The purchaser requires that the works should be completed not later than ................

   (b)  You are required to state in Part 1 of your tender the date by which you are prepared to undertake that the works will be completed ready to be put into commercial operation.

16 [*This paragraph to be included if nominated sub-contracts are involved.*]

You are required to quote in sub-section B of Part 2 of the Form of Tender your handling fee expressed as a percentage of each of the sums shown therein. The items shown will be the subject of nominated sub-contracts for which the purchaser will invite tenders from a list of contractors to be agreed with the successful plant contractor in accordance with condition ................ of the condition of contract. The successful plant contractor will also be responsible for preparing in conjunction with the purchaser's engineer the specification for sub-contracts listed in section B of Part 2 of the Form of Tender.

17 The following drawings and diagrams are enclosed to illustrate the requirements set out in the specification attached:

      TITLE      DRAWING AND DIAGRAM NUMBER

18 Your attention is drawn to the following [*insert here details of any particular requirements on safety – for example, compliance with works safety rules, prohibition on use of flame cutting apparatus, etc.*]

19 [*Include if necessary:*]
Tenderers are asked to note particularly that they should include in their tender for any overtime/weekend working caused by the need for breaking into existing structures, joining up to existing circuits, and so on.

20 (a) Notwithstanding the purchaser's right to reject any tender that is non-compliant, the purchaser reserves the right to seek further written clarification from the tenderer on any matter related to the tender.

   (b) Requests for clarification will be issued to the tenderer in writing, they are to be signed and returned by the tenderer and such clarification will be considered as part of the tender.

   (c) The purchaser also reserves the right to discuss the optimization of the preferred tenderer's proposals. The conclusion of such discussions will be treated in the same way as clarifications.

   [*This paragraph assumes in tenders subject to the EC Procurement and the Utilities Directives that the purchaser has selected the negotiated procedure. If the purchaser has selected the restricted procedure then only sub-paragraphs (a) and (b) could be included.*]

# Appendix 2  Bid desirability questionnaire

*A Marketing*

1 Does the tender fall within the main stream of
the company's activities or is it only peripheral?  10

2 How does the tender fit in with the company's
plans for market development or retention in
relation to the following factors:
  (a)   territory
  (b)   the particular customer
  (c)   the product(s) to be offered                 10
  (d)   the company's competitors?

3 What is the company's existing order book for
the product(s) concerned and what percentage of
the sales budget is covered by firm orders?         10

4 What alternative opportunities exist now or will
do so within the period covered by the tender for
the use of the same capacity?                        10

5 Of the balance of the sales budget uncovered by
firm order what are the chances of obtaining
other business on no less favourable terms?         10
                                                    ――  50

*B Production*

6  Would the contract if secured require any special facilities, e.g. special tooling, or involve the production of special parts or the use of non-standard components?                                     10

7  Would securing the contract impose any significant strain on production resources in terms of machines, labour inspection and test facilities, etc?                                                     10

8  What would be the effect of *not* securing the contract on:                                                10
   (a)  retention of staff/labour
   (b)  unrecovered overheads or adverse shop variances?

9  Has the product been manufactured before? If so, is it responsive to the customer's specification or are there risks in meeting mandatory requirements? If not, what degree of confidence exists in the ability of the product to meet such requirements?                                                20

                                                    — 50

*C Financial*

10  Is the anticipated cash flow positive or negative?  10

11  Are there any risks foreseen in relation to:
    (a)  cost escalation
    (b)  currency exchange rates                    } 15
    (c)  customer's financial stability?

12  Is the anticipated profit contribution as a minimum in line with the unit's planned target either overall or for that product line/market?  25

                                                    — 50

*D Contractual*

13  Will any contract be based on the company's or customer's terms?                                        —

14  Are there any contractual risks foreseen in relation to:

(a) penalty for delay
(b) warranty
(c) consequential damages
(d) inspection and testing requirements
(e) Inability to obtain truly independent decisions ⎫ 50
   on any disputes
(f) termination either for default or customer
   convenience
(g) performance guarantees?

50
───
200
───

# Appendix 3   Checklist of factors to be considered in comparing party and competitors

Price
Currency in which bidder is willing to accept payment
Willingness to accept payment under barter arrangement
Facilities for reciprocal trading
Ability to make off-set arrangements
Credit terms which can be offered
Availability of government to government loan
Delivery including reputation for keeping delivery promises
Risk of territory in which manufacture taking place being subject
   to industrial disputes
Conformity with mandatory specifications
Reliability of product
Quality of product
Ease of maintenance and level of running costs
Standardization with purchaser's existing plant/system
Ability to comply with performance guarantees
Design and technical merit of product
Capability of plant/system for expansion to meet purchaser's
   future requirements
After-sales service

Availability and price of spares

Willingness to accept purchaser's commercial terms of contract

Reputation on commercial negotiations of being 'hard' or 'soft'

Willingness to employ local labour

Willingness to manufacture locally

Willingness to make local investment

Local political connections

Location of manufacturing plant in development area or area of high unemployment

Quality of technical and managerial staff

Customer preference for company or product based on prior contact or knowledge

# Appendix 4   Questionnaire for site visits

---

*1.0   Proposed location of works*

1.1  (a)  Country . . . . . . . . . . . State or Province . . . . . . . . . .
     (b)  City or town nearest proposed site . . . . . . . . . . . . . . .
     (c)  Distance of site from city or town . . . . . . . . . . . . . . . .
          (include location map if available)
     (d)  If site owned or chosen give shape as:
          Length . . . . . . . . . . . . . . . Width . . . . . . . . . . . . . . .
          Total acreage . . . . . . . . . . . . . . . . . . . . . . . . . . . . . .
          Additional adjacent area available . . . . . . . . . . . . . . .
     (e)  Is there adjacent area available? . . . . . . . . . . . . . . . .
          State size . . . . . . . . . . . . . . . . . . . . . . . . . . . . . . . .
     (f)  Topography of site (level, rolling, steep etc.) . . . . . . . .
          . . . . . . . . . . . . . . . . . . . . . . . . . . . . . . . . . . . . . . .
     (g)  Drainage (describe) . . . . . . . . . . . . . . . . . . . . . . . . . .
     (h)  Are there any local codes governing construction? If
          so, a copy of the code should be obtained.

1.2  Foundations

*Soil characteristics*

Test results
Boring samples
Site geology
Access road soil bearing capacity
Soil analysis
Depth to water table (average)
Depth to rock (average)
Vegetation (type and density)
Obstructions above or below ground

1.3    Geographical considerations

Access to site
    Nearest national airport
    Nearest international airport
    Nearest rail head                        Max lift wt.
    Nearest ports                            Max lift wt.
    Nearest main roads
    Condition        Width                   Tonnes/Axle/limit
    Weight limitation Site to port           Width tonnes/Axle
                     Site to rail head Width tonnes/Axle
                     Site to airport
                     (national)
                     Site to airport
                     (international)

Bridge limitations:
    Site to ports
    Site to rail heads
Railway limitations: (truck capacity)
Accessible port with heaviest lift          Max lift wt.
    Site map
    Distance from switchyard intended
    site to nearest habitation              Metres
    Telephone and telex communication
    facilities available at site

1.4  Atmospheric conditions

Altitude above sea level:

*Annual temperature:*  Maxima:
Minima:
Average (design)

    *Monthly*  Maxima:    Daily max.
Minima:    Daily min.
Averages (design)

Relative humidities:

*Yearly*  Maxima:
Minima:
Averages (design)

*Monthly*  Maxima:
Minima
Averages (design)

Barometric pressure  Max.
Min.
Average (design)

Percentage sunshine days/average (design)
Annum – Average

Wind velocities—  Max vel:    Direction
Min vel:    Direction
Average vel:    Direction

Predominant direction of wind

Dust content

Unusual conditions, tornadoes,
cyclones, flood, earthquakes etc.

2.0  *Water supply*

(a)  Available quantity . . . . . . . . . if limited state min . . .
(b)  Source (as rivers, lakes, reservoirs, wells etc.) . . . . . . .
(c)  Distance from intake to plant site . . . . . . . . . . . . . . . .
(d)  Is there sufficient head for gravity flow of water to
works or must a pumping station be provided? . . . . .
. . . . . . . . . . . . . . . . . . . . . . . . . . . . . . . . . . . . . . . . .

(e)   Would supply be constant the year round? .........
(f)   If seasonal, state quantity fluctuations .... Min. ....Max.
(g)   Temperature at intake .........Min. .........Max.
(h)   General quality (as clear, cloudy, seasonably discoloured etc.) ...................................................
(i)   Would entire quantity of process water require filtration or treatment? ...........................
(j)   Or quenching quantity? ...........boiler feed water? ..................................................
(k)   Obtain water analysis if available, or send samples for analysis ........................................

### 3.0   *Power supply*

3.1   What is the power requirement:     KWH     Max.     KW
During construction?
During start-up?
For full production?

3.2   Is purchase power available at proposed site? ...........
(a)   Would it be available permanently? ...............
(b)   Is it a dependable source? .......................
(c)   Can a long-term contract be obtained?  ...........
(d)   What would be delivered current characteristics? .................................................
(e)   Would transformers and sub-station need to be supplied? ........................................
(f)   What losses would need to be allowed for?  ........
(g)   Would there be power factor penalties? ...........
(h)   What minimum charge for non-use? .............
(i)   On what basis purchased, i.e. per H.P. year, per KWH etc.  ..............................................
distance of delivery point
voltage at delivery point?
(j)   What agency is responsible?

### 4.0   *Gas supply*

Source of supply

Distance of delivery point
Calorific value
Analysis
What agency is responsible?

5.0  *Sewer effluents*

5.1  Foul sewer

    (a)   Location and size of main
    (b)   Invert elevations
    (c)   Owning agency
    (d)   Capacity of disposal plant
    (e)   Charges

5.2  Storm sewer

    (a)   Location and size of main
    (b)   Invert elevations
    (c)   Owning agency
    (d)   Outfall description
    (e)   Charges

5.3  Industrial effluent (liquid)

    (a)   Analysis of effluent
    (b)   Governing agency
    (c)   Schedule of requirements
    (d)   Treatment required

5.4  Industrial effluent (gaseous)

    (a)   Analysis
    (b)   Governing agency
    (c)   Schedule of requirements
    (d)   Treatment required
    (e)   Fume scrubbing

6.0  *Community data*

    (a)   Does a good, fair or poor labour market exist?
    (b)   How far removed from proposed plant site?

(c)   Type of transportation facilities to plant site?
(d)   Is housing available for additional personnel?
(e)   Predominant nationalities (or races) of community and percentage of each
(f)   School facilities      Universities
(g)   Churches and denominations
(h)   Hospitals (qualify as 'good' or 'fair')
(i)   Health and recreational facilities
(j)   Form of central and local government (state stability) Give names of prominent local officials
(k)   What is judicial system? Obtain copy of relevant codes

### 7.0  Shipping and transport

7.1  Landing costs
Harbour charges
Dockers charges
Duty (specify details if category subdivided)
Customs brokerage charges
Stamps and other duties
Customers brokerage (min. charges per consignment)

7.2  Transport to site charges

Parcels/boxes/crates less than 1000kgs per KG from harbour to site
As above but less than 10,000 KGS
As above but more than 10,000 KGS and less than 100,000 KGS
As above but over 100,000 KGS (to max. carrying capacity)
Are local trucking facilities available?

### 8.0  Construction requirements

8.1  Is there a good labour market available?  . . . . . . . . . . . . . .

8.2  (a)   If not, where is closest labour available? . . . . . . . . . . .
     (b)   Would transportation have to be paid for?  . . . . . . . . .
           How best arranged

Living cost · · · · · · · · · · · · · · · · · · · · · · · · · · · · · · · · · ·

8.3 (a) Are living quarters for construction crew available?
· · · · · · · · · · · · · · · · · · · · · · · · · · · · · · · · · · · · · · · · · ·

(b) Would temporary construction camps be required?
If so, is site available? · · · · · · · · · · · .Facilities needed
and available

(c) Feeding facilities · · · · · · · · Supply stores? · · · · · · · · ·

8.4 (a) What construction equipment is available (cranes, hoisting engines, concrete mixers, excavators, small tools etc)? State whether rental or sales basis · · · · · · ·
· · · · · · · · · · · · · · · · · · · · · · · · · · · · · · · · · · · · · · · · · ·

8.5 Are there competent local construction contractors who could undertake part or all of the work? Under what forms of contract do they normally operate? · · · · · · · · · · · · · · · ·
Are they willing to put up bonds?
*Prevailing rates of pay*
(a) *Carpenters* · · · · · · · · · · · · (b) *Bricklayers.* · · · · · · · · · · ·
(c) *Masons* · · · · · · · · · · · · · · · · · · · · · · · · · · · · · · · · · · · · ·
(d) *Steelworkers* · · · · · · · · · · · (e) *Riggers.* · · · · · · · · · · · · ·
(f) *Mechanics* · · · · · · · · · · · · · · · · · · · · · · · · · · · · · · · · · · ·
(g) *Foremen* · · · · · · · · · · · (h) *Skilled labour.* · · · · · · · · · · ·
(i) *Unskilled* · · · · · · · · · · · · · · · · · · · · · · · · · · · · · · · · · · ·

8.6 If labour unions are strongly organized, obtain current labour and trades schedule. Give particulars of union organization and with whom negotiations would need to be undertaken.

8.7 Obtain delivered to site prices, and location of sources available on:
(a) Cement · · · · · · · · · · · · · (b) Sand · · · · · · · · · · · · · · ·
(c) Gravel · · · · · · · · · · · · · (d) Lime · · · · · · · · · · · · · ·
(e) Brick or building tile · · · · · · · · · · · · · · · · · · · · · · · ·
(f) Lumber (all sizes) rough · · · · · · · · dressed · · · · · · · ·
(g) Millwork (as sash, doors etc.)
(h) Reinforcing steel · · · · · · · · · · · · · · · · · · · · · · · · · · · ·
Structural steel · · · · · · · · · · · · · · · · · · · · · · · · · · · · · ·

### 9.0  *Legal/commercial factors*

| | |
|---|---|
| Normal working week | Hours |
| Annual holidays | Days |

Normal working day                    am to        pm

Normal lunch interval                    to        hrs

Union or accepted standard skilled worker        per hour
(local currency)
      Semi-skilled
      Unskilled                         per hour

*(per hour entries:)*

Bilingual secretary      (local)        per hour
                                       per hour
Social security contributions
Health insurance                      % of above
Length of service indemnity
13th month salary
Holiday pay
Termination
Undue cause redundancy
Other on-costs (please specify)

# Index